Differential Games

Theory and Methods for Solving Game Problems with Singular Surfaces

Joseph Lewin

Differential Games

Theory and Methods for Solving Game Problems with Singular Surfaces

With 40 Figures

Springer-Verlag
London Berlin Heidelberg New York
Paris Tokyo Hong Kong
Barcelona Budapest

Joseph Lewin
Faculty of Aerospace Engineering, Technion-Israel Institute of Technology, Haifa 32000, Israel

ISBN 3-540-19841-5 Springer-Verlag Berlin Heidelberg New York
ISBN 0-387-19841-5 Springer-Verlag New York Berlin Heidelberg

Printed in Great Britain

Typesetting: Camera ready by author
Printed at the Alden Press, Oxford. Bound by Green Street Bindery, Oxford
69/3830-543210 Printed on acid-free paper

To my wife, Sara,

whose love, encouragement and support made the completion of this work possible.

List of Figures

Contents

Preface

Background and History

The name *differential games* hardly hints what is it all about and what are the expected benefits. (The term *game*, for example, when used as a key word for a search in literature, yields beside material about social and educational games also matters like game-animals, war-games etc.).

The theory of *differential games* is a discipline of applied science and as such we can *advertize* it by its potential applications as well as by its theoretical merits. Let us refer to both.

The theory of *Differential games* was developed as a means for modelling *conflicts*. In *real life* we have many kinds of conflict situations. Economic processes, labor-employers relations and even the issue of ecological equilibrium, are all influenced by conflicting interests. Military applications, obviously, provided the strongest stimulation to the development of the theory.
After World-War 2 weapons and counter-weapons became more sophisticated and expensive and a more appropriate modelling of problems characterized by conflict of interests was needed. Let us consider the problem of air-combat as an example. Even at the present, the result of an air combat encounter highly depends on the skills of the individual pilots. It is not unlikely that the side that will be the first to apply decision procedures derived from the analysis of a good game-model for air combat may gain a crucial advantage!

It is no wonder that the founder of the theory of differential games, *Ruphus Isaacs*[1], did his monumental work in the Rand corporation with an ori-

---

[1] One must appreciate the contribution and genius of *R. Isaacs*. We recall that many of the important results of control theory where not known and where found by Isaacs simultaneously and in a more general way. (See the famous review written by *Y. C. Ho* on Isaacs' book [Ho65]).

entation towards modelling and solving military motivated pursuit-evasion problems.

Isaacs discarded the reliance on simulations (*war games*) because they do not provide a sensible mean for *synthesizing* candidate optimal decision policies. He realized that the framework of existing theories could not accommodate the main feature of a conflict problem, namely the the fact that there exist at least two independent decision makers with conflicting objectives. He also noticed that the sequential structure of *classical game theory*, though it provided a setting for players with conflicting objectives, was not appropriate enough for modelling real-life problems (where typically one does not have to wait for his turn to make a decision as in a social game).
The theory of *differential games* is a blending of the notions of *control theory* with the decision structures and solution concepts of *classical game theory*. A differential game model cannot hence be considered as a *double* control problem. In general we can *reduce* a differential game model to a *control* model if we assume that only one player is active and the other is not.
Indeed this means that the theory of differential games includes the results of the theory of optimal controls as special cases!

The mathematical aspects of the theory of differential games attracted theoretical mathematicians almost immediately. *J.Danskin*, *L. D. Berkovitz*, *W. Flemming*, *A. Friedman*, *L. Pontriyagin*, *N. N. Krasovski. Petrosyan*, *P. L. Lions*, and *M. Bardi* are but few names.
The theoretical analysis of existence and optimality of solutions of differential games involves the mathematical theories of functional analysis and of partial and ordinary differential equations.

Isaacs was haunted by nontrivial pursuit evasion problems such as the *Homicidal Chauffeur* problem which he probably considered to be the most simplified model for a case in air combat where a slow, but more maneuvering airplane, is pursued by a faster and less maneuverable craft:

In an infinite planar parking lot a circular car (designated P) strives to knock a slower but more maneuverable pedestrian (designated E).

Surprisingly this is not a simple problem to solve. The *map* of outcomes of the optimal tactics is not smooth and it involves singularities reflecting the fact (which do not surprise those who study the history of actual conflicts) that at certain instances optimality is associated with circumstances where small deviations from the optimal tactics may cost the deviator a very high penalty.

Unlike some pure mathematicians who do not mind to use trivial examples to substantiate their theory, applied scientists should not avoid the singular phenomena but rather try to follow Isaacs and find means to device solution-techniques that accommodate them. Most outstanding among those who followed Isaacs, is with no doubt, *J. V. Breakwell.* His works, and later the works of his students (especially *P. Bernhard* but also *A. Merz* and myself), contributed to the solution of many problems and to the further development of the analysis of the phenomena of *singular surfaces* in differential games.

R. Isaacs, being an applied scientist, paid less attention to the mathematical rigor of his arguments and preferred geometrical approachs. No wonder that some pure mathematicians did not accept Isaacs' techniques for constructing singular surfaces in differential games for a long time. Only recently, as some of the singular phenomena can be discussed in the context of the newly introduced notion of *viscosity solutions* to the *Hamilton-Jacobi-Bellmann-Isaacs* equation this attitude changes.

Prospects

Differential games is, by far, the most appropriate discipline of applied science for modelling conflict problems of real life in an analytical fashion.
Ever since Isaacs book was published we evidenced periodical surges of enthusiasm concerning the prospects of obtaining a major practical achievement by applying the techniques of the theory of differential games to an important problem from real life. One wave of enthusiasm occurred when *A. Bryson* , *Y. C. Ho*, and *S. Baron* presented [ABB65] a model within which they claimed that the famous *proportional navigation guidance rule* (which was originally designed only by engineering considerations) is a realization of an optimal game strategy.
Soon people realized that the actual construction of solutions to high dimensional differential games is laborious and difficult.
The fast improvement of computers in the 80's brought a renewed research effort, led by J. Shinar and his collaborators, to incorporate solutions of simplified subgames in a design of a *pilot advisory system* for future air combat.
A new interest in the theory of differential games, and this time among researchers in control theory, aroused when the linkage between the notion of *robust optimal control* and Linear-Quadratic differential games was applied in *controller design*. (See [BB91] and [SR91]).

Purpose of the Book

This book is particularly aimed to fulfil the needs of those who want to study the theory of differential games in order to use it for applications.
The book is focussed on the techniques and considerations regarding the construction of solutions of differential games with singular surfaces. It contains results derived in the period after the latest edition of Isaacs' book was published (1975).
This book is based on my experience in teaching the graduate course on the subject of differential games in the *Technion* for many years.
Many exercises and worked examples were included so that the reader can practice and gain actual experience.
The book may serve as a text for a graduate course or for self study.

Methodology

When I wrote the book I faced some dilemmas about the scope and the style.

First I had to decide about the mathematical style.
If one wants to write proofs in a pure rigorous style he can not avoid discussing many mathematical details and using very cumbersome notations. Readers with an engineering background may feel very uneasy and discouraged encountering too many unfamiliar mathematical notions.
On the other hand a certain degree of mathematical rigor is essential so that the reader will understand the limitations of the theory and thus avoid mistakes caused by misusing the techniques.
I decided to prefer the use of geometric interpretations of optimization theory for clarifying the mathematical ideas rather than to use a very formal style.
I also tolerated, in cases when it contributed to clarity and caused no ambiguity, some abuse of the notations.

The second issue was to decide about the *scope* of the text. I preferred to concentrate on *two-players zero-sum perfect-information differential-games* (which is the backbone of the theory), rather than spread the discussions over many topics, believing that a more solid knowledge of the main theory will enable the reader to follow other directions[2] more easily.

Many worked examples are included to illuminate specific issues and to

[2] In the literature one may find papers on differential games with many players, with cooperative preferences and with imperfect, noisy, or delayed information.

demonstrate an actual use of the procedures. At the end of most chapters there is a section with problems to be worked out by the reader.
The examples and the problems are mostly nontrivial but their resolution do not involve numerical procedures. The reader can thus grasp the flavor of the art of solving differential games with less effort.
The resolution of some research problems may involve a lot of numerical computations. Special care should be taken in identifying the singular surfaces since the current numerical procedures may fail in doing that.
A discussion of aspects of the numerical computations in the course of construction of solutions to differential games is outside the scope of this text.

About Reading the Book

Now a brief outline of the chapters will be given.
Chapters 1 to 4 constitute the preliminary part of the book where notions are defined and concepts are explained.
Chapters 5 to 10 contain the theoretical background for the methods of construction of solutions to differential games. After reading chapter 7 and studying the examples given there, the reader will be able to solve simple games.
Chapters 11 to 15 each discuss a specific class of singular surfaces.

Chapter 1 is devoted to a *preview example.* It is an easy example of a differential game and no prior knowledge is required. It is recommended that a reader new to the subject shall read this first chapter first.
Chapter 2 presents the vocabulary for differential games. Some of the notions resemble those of *control theory* and others resemble *game theory* but the ensemble is particular to the theory of differential games.
Chapter 3 discusses the issue of *solution concept.* Two solution concepts are explained. The classical Isaacs' concept and an extended concept that can accommodate the singular phenomena.
Chapter 4 gives a preliminary discussion of the important issue of *semipermability of surfaces* which is of value in a much wider context than the theory of differential games where it originated.
Chapter 5 is on *necessary conditions.* We present the reader with a systematic derivation, based on the definitions and the notions of the solution concept, of the regular necessary conditions that candidate solutions must fulfil. The chapter includes a discussion of a *viscosity form of Isaacs Equations.*
Chapter 6 refers to the subject of sufficiency and discusses conditions for validity of partial solutions. *Chapter 7* deals with the *regular* construction methods that suffice for solving differential games which do not have *singular phenomena.* The chapter also includes a discussion of the basic theory

of *Linear Quadratic differential games* and of its relevance to the area of *robust controller design*.
Chapter 8 deals with the construction of semipermeable surfaces. The methods of differential games are applied in a much broader context.
Chapter 9 gives a descriptive preview of the singular surfaces that we may encounter in the solution of differential games. We make an analogy between the *value map* in differential games and *topography*.
Chapter 10 contains a derivation of the necessary conditions that have to be met at optimal arcs that lie on singular surfaces. The unified theory covers the subject of the junction between the regular and singular parts of an optimal trajectory. The issue of ϵ-optimal strategies is discussed.
Chapter 11 deals with *dispersal surfaces*. At a dispersal surface one player has more then one optimal choice. Optimal trajectories leave to both sides of a dispersal surface.
Chapter 12 treats singular arcs of *safe contact*. Such a singular arc follows adjacent to the *target-set* or to a locus of a *state constraint*.
Chapter 13 analyzes *universal surfaces* and *focal surfaces*. Optimal trajectories arrive at both sides of those surfaces and proceed adjacent to them.
Chapter 14 discusses *corner surfaces*. Optimal trajectories arrive at one side of a corner surface and may either proceed along a singular arc adjacent to it or switch to the other side regularly.
The subject of *Chapter 15* is the *envelope barrier*. Here argument similar to those used in the construction of singular surfaces in differential games are used in a derivation of a method to construct a particular kind of *semipermeable surfaces*.

1

A Preview Example

1.1 Introduction

The theory of differential games is used to model dynamic conflicts. Some of the terms that we use for the analysis of differential games resemble the terms used for control theory and other resemble those used in game theory. We shall see that the theory of differential games may be viewed as a hybrid of those theories.

Before we go into the detailed analysis it may be advantageous to discuss a very simple old example as a preview. (Recall that it is much easier to fit the pieces of a jig saw puzzle when we know how the complete picture will look).

1.2 A Simple Differential Game

The *game* takes place in a *game set* that consists of the upper half of a plane. Two players, *player P* and *player E* are concerned with the control of the motion of a point in the plane. The location of the point (its *state*) at the time t is represented by a vector $\mathbf{x}(t)$ with components $[x(t), y(t)]$. The motion is described by the following *equations of motion*:

$$\begin{aligned} \dot{x} &= Av + B\sin u, \qquad |v| \leq 1 \\ \dot{y} &= -1 + B\cos u \end{aligned} \tag{1.1}$$

Where A and B are constants and $0 < B < A < 1$.

At each instant of time *player P* chooses a value for the *control* u and *player* E chooses a value for the *control* v. Notice that only choices of v that satisfy $|v| \leq 1$ are considered to be *admissible controls* for *player E*.

A *play* may start at any initial *state* (x, y) in the *game set* (namely with $y \geq 0$). As the *play* proceeds the motion of the point (x, y) describes a *trajectory* in the *game set*. A play *terminates* when the *trajectory* of the *play* crosses the x axis at some *terminal state* $(x_f, 0)$.

An *outcome* is associated with each *terminating play*. It is defined by the value of a given function G at the *terminal state* of the *play*. Let us assume that

$$G(x, y) = x. \tag{1.2}$$

Player P and *player E* have totally conflicting *preferences* regarding the *outcome*. *Player P* wants to minimize it while *player E* wants it to become

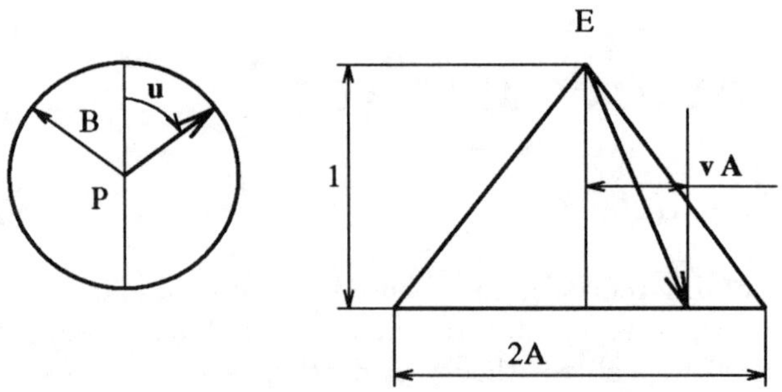

FIGURE 1.1. The vectograms

maximal.
Let us consider a *play* that starts at some (x, y).
Can *player* P find a rule (*strategy*) for choosing the control u that guarantees him that the *outcome* will not exceed some value $J_p(x, y)$?
Does *player* E have a *strategy* that guarantees him an *outcome* of at least $J_e(x, y)$ as well?
Is it possible that $J_p(x, y) = J_e(x, y)$?

1.3 Preliminary Analysis

We may view equations (1.1) of section 1.2 as equations of motion for the components of a *velocity vector* $\dot{\mathbf{x}} = \mathbf{f}(\mathbf{x}, u, v)$.
They describe the motion of the *state vector* $\mathbf{x} = x\hat{\mathbf{e}}_x + y\hat{\mathbf{e}}_y$ where $\hat{\mathbf{e}}_x$ and $\hat{\mathbf{e}}_y$ are *unit vectors* in the directions of the x and y axis respectively:

$$\dot{\mathbf{x}} = (Av + B\sin u)\hat{\mathbf{e}}_x + (B\cos u - 1)\hat{\mathbf{e}}_y$$

An alternative componentization of the velocity vector $\dot{\mathbf{x}}$ is possible in our case:

$$\dot{\mathbf{x}} = ((B\sin u)\hat{\mathbf{e}}_x + (B\cos u)\hat{\mathbf{e}}_y) + (Av\hat{\mathbf{e}}_x - \hat{\mathbf{e}}_y).$$

Here the first component (the first rectangular bracket) is dominated by *player* P and the second by *player* E.
We can give a geometric description for the range of each of those components (over the set of possible choices of controls). We call those geometric descriptions *P-vectogram* and *E-vectogram* (see Figure 1.1).
From equation (1.1) we can deduce that $\dot{y} < 0$ all the time. This means that the *trajectory* of every *play* intersects the x axis in finite time. In other words, all the *plays* in this game *terminate*.

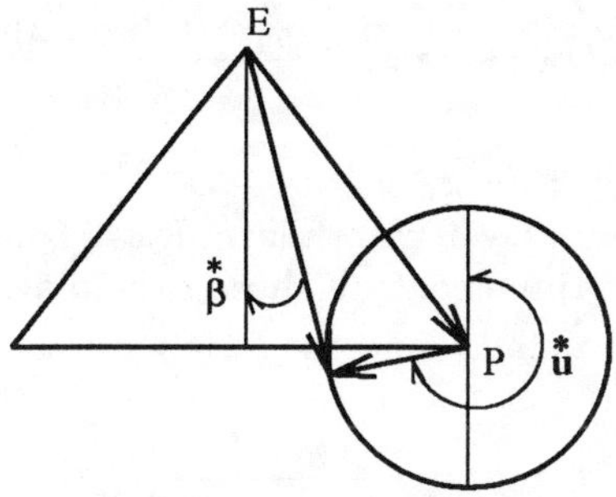

FIGURE 1.2. The combined vectogram

Notice also that *player E* dominates the sign of the x component of the *velocity vector* while *player P* governs its y component.

1.4 A Heuristic Solution

Let us now try to figure out, for each player, if there exists a rule (*strategy*) for choosing his control at each instance that provides him a best possible guaranteed outcome.
Player E wants to maximize x_f. His *strategy* will certainly be to choose $\overset{*}{v}= 1$ everywhere. (We use the $*$ superscript to designate that this choice of v is a *candidate optimal control* for player E).
The *strategy* of player P, who strives for minimal x_f, should be to choose a control $\overset{*}{u}$ that minimizes the inclination β of the *trajectory* with respect to the vertical direction. The *candidate optimal control* $\overset{*}{u}$ and the corresponding $\overset{*}{\beta}$ can be found from the geometry of the *vectograms*, (see Figure 1.2).
Let us consider a *play* that starts at (x_0, y_0). Let $\overset{*}{u}_0$ be the *optimal* choice of control by player P there. Player P can thus guarantee that the duration of the play will not exceed $\frac{y_0}{1-B\cos\overset{*}{u}}$ and that $x_f \leq x_0 + \frac{y_0(a+B\sin\overset{*}{u})}{1-B\cos\overset{*}{u}}$.
For a *play* that starts at (x_0, y_0) player P can therefore guarantee himself an outcome $J_P(x_0, y_0)$ which is at most

$$J_P(x_0, y_0) \leq x_0 + \frac{y_0(a + B\sin\overset{*}{u})}{1 - B\cos\overset{*}{u}}.$$

Similarly,by choosing $\overset{*}{v}= 1$, player E can guarantee that

$$\overset{*}{\beta}\leq \arctan\frac{(a + B\sin\overset{*}{u})}{1 - B\cos\overset{*}{u}},$$

and that the outcome $J_E(x_0, y_0)$ will be at least

$$J_E(x_0, y_0) \geq x_0 + \frac{y_0(a + B \sin \overset{*}{u})}{1 - B \cos \overset{*}{u}}.$$

Notice that $J_E(x_0, y_0) = J_P(x_0, y_0)$.
It is evident that neither player P nor player E can benefit from a unilateral deviation from their optimal controls during the play.
The *optimal strategies* $\overset{*}{u}$ and $\overset{*}{v}$ and the function:

$$J(x, y) = x + \frac{y(a + B \sin \overset{*}{u})}{1 - B \cos \overset{*}{u}},$$

are considered to be a solution of this game because they provide the players with guidance how to guarantee an optimal outcome $J(x, y)$ for plays that start at any possible state (x, y).
We call $J(x, y)$ the *value function* of this game.
If both players realize their *optimal strategies* and choose the optimal controls $(\overset{*}{u}, \overset{*}{v})$ the corresponding *optimal trajectories* will be straight lines. Notice that along an *optimal trajectory* the *value* $J(x, y)$ is constant. We say that in our game the *optimal trajectories* are also *iso value surfaces*.

1.5 Problems

Problem 1.5.1

Consider the Preview Example.

1. Evaluate explicit expressions for $\overset{*}{u}$, sin $\overset{*}{\beta}$, and $J(x, y)$ as functions of the parameters A and B.

2. Verify that the lines $J(x, y) = c$ have the following property: At each point *player* P can prevent the play from crossing the line in the direction of $\nabla J(x, y)$ and that *player* E can prevent crossing in the direction of $-\nabla J(x, y)$.
   Does this property depend on the function G?

3. Draw a map of *iso-value lines* that correspond to the following cases:
   (a)
   $$G(x) = (x + 1)^2.$$
   (b)
   $$G(x) = -(x + 1)^2.$$

(c)

$$G(x) = \min[(x+2)^2, (x-2)^2].$$

(d)

$$G(x) = \begin{cases} x^2, & \text{for } x \leq o \\ x^{-1}, & \text{for } x > 0. \end{cases}$$

(e) Is it optimal for player P to play a constant control in those cases? compare to the example in the text.

2

The Vocabulary For Differential Games

2.1 Introduction

Some readers may be applications oriented. They expect the theory of differential games to provide them with tools that will enable them to model and analyze real life problems in a simple and meaningful way. The motivation of *R. Isaacs* [Isa75], for example, in formulating the *Homicidal Chauffeur* game was to use it as a simplified model for a nontrivial case of air-combat where a slow maneuverable airplane is pursued by a faster but less maneuverable aircraft.
This group of readers obviously prefer that the formulation will be both clear and broad so that the features of a wider class of real life problems could be easily and adequately modelled.
Mathematicians, on the other hand, consider differential games as a topic related to the theory of optimization and might be more interested in the theoretical aspects of the subject. They want the theory to be stated rigorously. The consequences of this requirement depend on our approach. If we look for a more general theory the formulation will become too complex and the scope of applications will be limited to fewer classes of models for which theoretical solutions exist. If, on the other hand, we simplify the formulation by using more restrictive mathematical definitions, we may be limited to a narrower class of features of real life problems that can be adequately modelled by the theory.
We shall try to keep the right balance between the two approaches in contents as well as in style.
In this chapter we shall discuss the definitions in the order arising from the procedure with which we model a case of a real life conflict as a differential game:

- The *state vector* and the *game set*.
- The *equations of motion*, (velocity vector and admissible controls).
- *Terminating situations*, (target-set and state constraints).
- *Plays*, (terminating and nonterminating).
- *Outcomes*, (preferences of the players).
- *Information constraints*.

- *Strategies*, (admissible and nonadmissible).
- Strategies that guarantee termination or nontermination.

2.2 The State Vector and the Game-Set

Suppose that we want to convert a case of a real life conflict to a differential *game model*. The first step in the process is to decide what is the set of variables that characterize the conflict. We then consider this set of variables as the components of a *vector* $\mathbf{x}$ which we call the *state vector*. We refer to the number of the components of the state vector as the *dimension* of the differential game problem.
The set of all possible states is defined as the *game-set* (which we shall designate by $\mathcal{S}$). In mathematical terms the *game-set* $\mathcal{S}$ is the range of the *vector* $\mathbf{x}$ and is a subset of a n-dimensional Euclidean space which we shall call the *state space* (designated as $\mathcal{X}$).

In general, the state vector representation for a specific real life problem is not unique!
We may choose the dimension and the particular componentization of the state vector from practical considerations. We may augment the state space by adjoining *time* t as an additional component of the state vector. We can take advantage of the fact that the physical features of a real life problem should obviously be invariant to the particular representation by investigating different formulations simultaneously.

Example 2.2.1 *[The Homicidal Chauffeur*[1] *]*

In an infinite planar parking lot a circular *car* P strives to knock a slower *pedestrian* E. The *car* has a radius β, a maximal speed of magnitude 1 and a minimal turn radius of magnitude 1. The *pedestrian* has a maximal speed of magnitude $\gamma < 1$ and has a capability to perform instantaneous turns.

Let us consider various state vector representations with respect to a *fixed* frame of reference OXY in the plane and to a *relative* reference frame Pxy attached to the *car* P (see Figure 2.1). We may use any of the following state vectors to represent the status of the system:

1. $\mathbf{x} = (X_p, Y_p, X_e, Y_e)$, or $\mathbf{R} = (\mathbf{R}_p, \mathbf{R}_e)$.
2. $\mathbf{x} = (\mathbf{r}, \varphi)$, where $\mathbf{r} = \mathbf{R}_p - \mathbf{R}_e$.

---

[1] The problem of *the Homicidal Chauffeur* was posed and analysed by R. Isaacs [Isa75]. Isaacs did not complete the solution of the problem. The complete solution was done by *A. Merz* in his Ph.D thesis and is described in [Mer71].

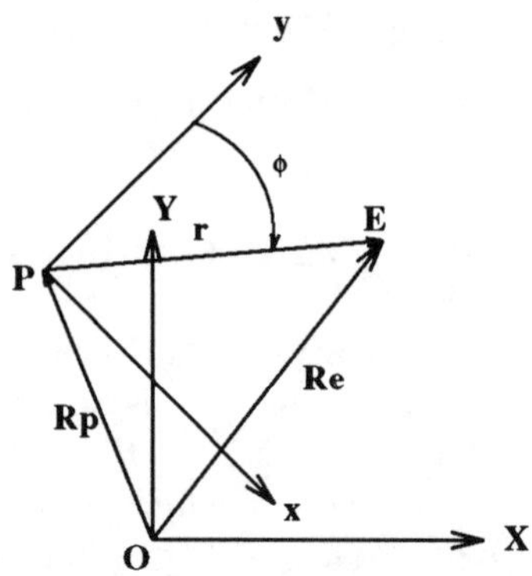

FIGURE 2.1. Reference frames in the Homicidal Chauffeur game

3. $\mathbf{x} = (x, y)$.
4. $\mathbf{x} = (r, \varphi)$, where $r = |\mathbf{R}_p - \mathbf{R}_e|$.

•

2.3 The Equations of Motion

When experts in political sciences analyze possible future *situations* in a real life conflict they usually discuss several *scenarios* that differ from each other due to the various decision patterns assumed for the rival sides.
In a differential game the rivals are called *players*. We call one player *player* P and the other *player* E[2]. The influence of the decisions of the players on the evolution of the state of the game is described mathematically by the *equations of motion*.

The equations of motion will usually have the following form:

$$\dot{\mathbf{x}} = \mathbf{f}(\mathbf{x}, \mathbf{u}, \mathbf{v}) \;, \quad \mathbf{u} \in \mathcal{U} \;, \quad \mathbf{v} \in \mathcal{V}.$$

The variables $\mathbf{u}$ and $\mathbf{v}$ represent the influence of the players on the *velocity vector* $\mathbf{f}$ of the game.
$\mathbf{u} = (u_1, \ldots, u_k)$ and $\mathbf{v} = (v_1, \ldots, v_m)$ are the *control vectors* chosen at each instance by player P and player E respectively.
The rules of the game may restrict $\mathbf{u}$ to belong to a prescribed set $\mathcal{U}$ and

[2]Those names originate from *pursuit evasion* problems where P stands for "*pursuer*" and E stands for "*evader*".

$\mathbf{v}$ to belong to a prescribed set $\mathcal{V}$[3]. $\mathcal{U}$ is the *set of admissible controls* for player P and $\mathcal{V}$ is the *set of admissible controls* for player E.
We characterize the mathematical properties of $\mathbf{f}(\mathbf{x}, \mathbf{u}, \mathbf{v})$ as follows:

1. The sets of *admissible controls* $\mathcal{U}$ and $\mathcal{V}$ are assumed to be convex sets. This is not a severe restriction. In real life problems the use of *chatter* between two values of controls is often feasible so if the set $\mathcal{U}$ is originally nonconvex it can be replaced by its *convex hull*. We further assume that the sets of *admissible controls* $\mathcal{U}$ and $\mathcal{V}$ do not depend on the *state* $\mathbf{x}$ as long as $\mathbf{x}$ is not on the boundary of the *game set*.

2. $\mathbf{f}(\cdot,\cdot,\cdot)$ is a single valued function, convex[4] in $\mathbf{u}$ and $\mathbf{v}$.
At any *state* $\mathbf{x}$ there corresponds a *unique velocity vector* to each pair of choices $(\mathbf{u}, \mathbf{v})$.

3. The range of $\mathbf{f}(\mathbf{x}, \mathbf{u}, \mathbf{v})$ is bounded over the domains of its independent variables.

Sometimes $\mathbf{f}(\mathbf{x}, \mathbf{u}, \mathbf{v})$ can be separated to two parts where one part is *dominated* by player P and the other is *dominated* by player E,

$$\mathbf{f}(\mathbf{x}, \mathbf{u}, \mathbf{v}) = \mathbf{f}_P(\mathbf{x}, \mathbf{u}) + \mathbf{f}_E(\mathbf{x}, \mathbf{v}). \tag{2.1}$$

A *velocity vector* $\mathbf{f}$ that has the property of equation (2.1) is called *separable*. When a velocity vector is separable we can use *vectograms* (see Section 1.3).

2.4 Termigation of a Differential Game

Some situations in a conflict in real life are considered as *end* or *kill*. Frequently such an *end* is associated with a certain contact between the combatants. Consider *bull-fighting* as an example. Let us examine the relative velocity between the *bull* and the *fighter* at an instance of *contact* and use it to distinguish between three types of *contacts*:

Separating contact: the *bull* and the *fighter* are moving *away* from each other.

Grazing contact: the *bull* and the *fighter* move *tangent* to the plane of contact.

[3] Those restrictions are referred to as *control constraints*.
[4] A function $g(u)$ is convex in u if $g[\varepsilon u_1 + (1-\varepsilon)u_2]$ is in the range of $g(u)$ for all $|\varepsilon| \leq 1$ and for all u_1 and u_2 in the domain of $g(u)$.

Penetrating contact: the *bull* and the *fighter* move *towards* each other.

Similar contact situations occur also in other *pursuit evasion* conflicts.
In a *bull fight*, as well as in many other pursuit evasion problems in real life, the rules state that only a *penetrating contact* is considered as a legitimate end situation of the fight.
We shall apply this rule to the formulation of differential game models too.
To do so we introduce two terms: *target-set* and *termination situation*.

The target set

The *target-set* $\mathcal{C}$ is a subset of the boundaries $\partial\mathcal{S}$ of the game-set $\mathcal{S}$. It consists of all states $\mathbf{x}$ at which the crossing of the boundaries is allowed by the rules of the game.

Termination situations

A *termination situation* is a triplet $(\mathbf{x}, \mathbf{u}, \mathbf{v})$ where the *state* $\mathbf{x}$ is on the *target set* and the *velocity vector* $\mathbf{f}(\mathbf{x}, \mathbf{u}, \mathbf{v})$ penetrates the boundary at $\mathbf{x}$ in a direction pointing *outside* the game set. Let $\mathbf{n}(\mathbf{x})$ be a vector normal to the boundary $\partial\mathcal{S}$ at $\mathbf{x}$ pointing *into* the game set. $(\mathbf{x}, \mathbf{u}, \mathbf{v})$ is a *termination situation* when

$$\mathbf{f}(\mathbf{x}, \mathbf{u}, \mathbf{v}) \cdot \mathbf{n}(\mathbf{x}) < 0. \tag{2.2}$$

We shall restrict ourselves to cases where the target-set $\mathcal{C}$ is a piece-wise smooth surface $C(\mathbf{x}) = 0$[5].

Example 2.4.1 (The Homicidal Chauffeur)

Consider the *Homicidal Chauffeur* problem (see Example 2.2.1) again. The *target-sets* for the various formulations are:

1. $(X_p - X_e)^2 + (Y_p - Y_e)^2 = \beta^2$, or $|\mathbf{R}_p - \mathbf{R}_e| = \beta$.
2. $|r| = \beta$.
3. $x^2 + y^2 = \beta^2$.
4. $r = \beta$.

Notice that here the *target-set* coincides with the boundary of the *game-set*.

●

[5] By definition $\nabla C(\mathbf{x})$ is aligned with $\mathbf{n}(\mathbf{x})$ so relation (2.2) is equivalent to:

$$\mathbf{f}(\mathbf{x}, \mathbf{u}, \mathbf{v}) \cdot \nabla C(\mathbf{x}) < 0. \tag{2.3}$$

Example 2.4.2 *[The problem of Lion and Man[6]]*

A *man*, armed with a gun that has an effective range of magnitude $\rho < 1$, chases a *lion*. The chase takes place in a circular arena with radius of magnitude 1. The speed of *lion*, v_L, is greater than the speed of v_M of *man*. Both *lion* and *man* are capable of instantaneous turns. *Lion* is clever and in spite of the fact that he is faster, he tries to evade *man*.

We may formulate this game in a 4-dimensional state space :

The state vector:

$$\mathbf{R} = (\mathbf{R}_M, \mathbf{R}_L),$$

The game-set:

$$\mathcal{S} \equiv \{\mathbf{R} \mid \ |\mathbf{R}_M| \leq 1, \quad |\mathbf{R}_L| \leq 1, \quad |\mathbf{R}_M - \mathbf{R}_L| \geq \rho\},$$

The target-set:

$$\mathcal{S} \equiv \{\mathbf{R} \mid \ |\mathbf{R}_M| \leq 1, \quad |\mathbf{R}_L| \leq 1, \quad |\mathbf{R}_M - \mathbf{R}_L| = \rho\},$$

In this case part of the boundary of the *game-set* is not in the *target-set*. This part constitutes a *state constraint* where both players are not allowed to cross.

•

2.5 Plays

When we discuss a situation of a conflict in real life we have to make a clear distinction between a *posterior* analysis (analysis of past events) and a *prior* analysis (analysis of future developments). The past events of a conflict are *history* and can be described chronically in a unique way. It is possible to give a time record of the past evolution of the system and of the decisions made by the rivals. The discussion of future developments in a conflict cannot be done in this manner. We cannot assume a unique future evolution anymore because we do not know the decisions of the rival ahead of time.
When we discuss differential games we have to do a similar distinction between *posterior description* and *prior analysis.*

A past evolution of the states of a system is given by a continuous function

[6] The problem of *Lion and Man* belongs to the class of problems called *pursuit of a faster evader* (See *J.V. Breakwell* in [Bre75]). The problem was treated by *J. Flynn* in [Fly73,Fly74], by *J. Lewin* in [Lew86] and by *J. V. Breakwell* in [Bre89].

of time $\mathbf{x}(\cdot)$ where $\mathbf{x}(t)$ is the *state* of the system at time t.
We assume that any past evolution $\mathbf{x}(\cdot)$ of the system can be characterized by a triplet

$$[\mathbf{x}_0, \mathbf{u}(\cdot), \mathbf{v}(\cdot)],$$

where $\mathbf{x}_0$ is the initial state and $\mathbf{u}(\cdot)$, $\mathbf{v}(\cdot)$ are time *histories* of the choices of controls $\mathbf{u}(t)$ and $\mathbf{u}(t)$ made by player P and player E[7].

A play
We call a triplet $[\mathbf{x}_0, \mathbf{u}(\cdot), \mathbf{v}(\cdot)]$ that characterizes a past evolution of the state of a system in a game a *play*.

Control histories
The functions $\mathbf{u}(t)$ and $\mathbf{u}(t)$ are called *control histories*.

Plays do not cross a *state constraint* because by their definition they are correlated to past evolutions of the *state* within the *game set*.

We distinguish between two types of *plays*.

A terminating play
In a *terminating play* a *termination situation* (see Section 2.4) is attained at a certain time t_f.

A nonterminating play
In a *nonterminating play* the system does not attain a *termination situation* (it may wander in the *game set* forever).

2.6 Outcomes

In a real life conflict problem we often have to choose between alternative decision policies. We compare the results associated with each alternative policy and we then choose the policy that is expected to yield the best outcome.
The key elements of this comparison process, using the terminology of the theory of *decision analysis*, are:

Events
A set of *evolutions* that correspond to the different decisions.

Results
A unique *result* that is associated to each *events* (decision theory does not require *results* to be numbers!).

---

[7] Notice that this assumption differs from an assumption that a unique solution $\mathbf{x}(t)$ of the ordinary differential equation $\dot{\mathbf{x}} = \mathbf{f}(\mathbf{x}, \mathbf{u}, \mathbf{v})$ from $\mathbf{x}_0$ exists for any given functions u and v.

Preferences

The *decision maker* is characterized by his *order of preference* over the set of *all* the possible *results*.

Let us now see what are the *events, results* and *preferences* in a game model regarding the decisions made by the players as they choose the controls. In a differential games the *evolutions* (**events**) that correspond to alternative pairs of decision histories are the corresponding *plays*.

To each play[8] we associate a unique *result* which we shall call *the outcome of the play*. We define the result associated with nonterminating and terminating plays as follows:

The result of a nonterminating play

The *result* of all the *nonterminating plays* will be the same. We denote it by Υ[9]. This result is not a number.

The result of a terminating play

Let $[\mathbf{x}_0, \mathbf{u}(\cdot), \mathbf{v}(\cdot)]$ be a *play* that terminates at the time t_f at the *state* $\mathbf{x}(t_f)$. The result associated with such a *terminating play* will be defined by the following *outcome functional*:

$$\pi[\mathbf{x}_0, \mathbf{u}(\cdot), \mathbf{v}(\cdot)] = \int_0^{t_f} L[\mathbf{x}(\tau), \mathbf{u}(\tau), \mathbf{v}(\tau)]\, d\tau + G[\mathbf{x}(t_f)].$$

The function $L(\mathbf{x}, \mathbf{u}, \mathbf{v})$ is called the *running cost function*. We shall assume that it has the same mathematical properties as the components of $\mathbf{f}(\mathbf{x}, \mathbf{u}, \mathbf{v})$. (See Section 2.3).

The function $G(\mathbf{x})$ is a single valued scalar function defined on the *Target Set* and is called the *terminal cost function*. We allow $G(\mathbf{x})$ and its derivatives to have a finite number of jump discontinuities.

Differential games are often classified according to the type of their *outcome functional*:

A Game of Kind

Sometimes, for some real life pursuit evasion problems, we only look for the boundaries of a *capture-zone*, the set of initial positions for which the pursuer can assure capture in finite time. In such cases we may define a differential game model where the outcome may attain only two values, Υ for a non-terminating play and some constant number for all terminating plays.

A case, like this, when $G(\mathbf{x}) = c$, (c is a constant), and $L(\mathbf{x}, \mathbf{u}, \mathbf{v}) = 0$

---

[8] Recall that we are obliged to define a result to *all* possible plays!

[9] We do not want to define it by ∞ or $-\infty$ as some authors do.

where the outcome functional of a differential game may assume only two values[10], was called by Isaacs a *Game of Kind*.

A Game of Degree

The case when the outcome functional of a differential game may assume a continuum of possible values was called by Isaacs a *Game of Degree*.

We may distinguish between the following two types of Games of Degree:

A Running cost game

When $G(x) = 0$ and $L(\mathbf{x}, \mathbf{u}, \mathbf{v}) \not\equiv 0$ we have the case of a *Running Cost Game*. Notice that if $L = 1$ the outcome of any terminating play is the *time to capture*. This case is sometimes called *pure pursuit evasion game*. The *Homicidal Chauffeur* game (see example 2.2.1) is an example for a *pure pursuit evasion game*.

A Terminal Cost Game

When $L(\mathbf{x}, \mathbf{u}, \mathbf{v}) = 0$ and $G(\mathbf{x})$ assumes a continuum of possible values it is a *terminal cost game*. The *preview example* (see Section 1.2) is a *terminal cost game*.

Lemma 2.6.1

Any n-dimensional game with a running cost term in the outcome functional can be reformulated as a $(n+1)$-dimensional *terminal cost game*.

Proof

We introduce the new variable $x_{n+1}(t)$ where

$$\dot{x}_{n+1} = L(\mathbf{x}, \mathbf{u}, \mathbf{v}),$$

$$x_{n+1}(t) = \int_0^t L[\mathbf{x}(\tau), \mathbf{u}(\tau), \mathbf{v}(\tau)]\, d\tau.$$

We can now adjoin x_{n+1} as the $(n+1)$th component to the the original *state vector* $\mathbf{x}$ and thus obtain a new *state vector* $\mathbf{z}$ and a new velocity vector $\dot{\mathbf{z}}$ that represent a terminal cost game in a $(n+1)$ dimensional game set:

$$\mathbf{z}(t) = \begin{pmatrix} \mathbf{x}(t) \\ x_{n+1}(t) \end{pmatrix} \quad \text{and} \quad \dot{\mathbf{z}} = \begin{pmatrix} \mathbf{f}(\mathbf{x}, \mathbf{u}, \mathbf{v}) \\ L(\mathbf{x}, \mathbf{u}, \mathbf{v}) \end{pmatrix}$$

The new *running cost function* $\Lambda(\mathbf{z}, \mathbf{u}, \mathbf{v})$ and the new *terminal cost function* $\Gamma(\mathbf{z})$ are:

$$\Lambda(\mathbf{z}, \mathbf{u}, \mathbf{v}) = 0 \quad \text{and} \quad \Gamma(\mathbf{z}) = x_{n+1}$$

•

---

[10] Or even a finite number of different values.

Let m and n be different results corresponding to two different plays.
If player P prefers the *result* m over the *result* n we shall say that the relation denoted by $m \overset{p}{\succ} n$ is an *order of preference* of player P over those results. We recall from *decision theory* that an *order of preference* is, in the mathematical sense, a *partial order*. It is required that a fixed[11] unique order relation will exist over the set of all possible *results* of plays and that it will be *transitive*:

$$a \overset{P}{\succeq} b \quad \text{and} \quad b \overset{P}{\succeq} c \quad \text{implies} \quad a \overset{P}{\succeq} c.$$

This set may in general include the non-number result Υ that corresponds to *nonterminating plays* beside numerical results that correspond to *outcomes* of *terminating plays*. Never the less *order of preference* can be defined over such 'mixed' sets too.

In a conflict, the results are influenced by more than one *decision maker*. Different orders of preference may yield different game models.
In this book we shall restrict ourselves to problems of *pure conflict*[12]. In a *pure conflict* the *orders of preference* of player P and player E are the reverse of each other:

$$\text{if} \quad m \overset{P}{\succ} n \quad \text{then} \quad n \overset{E}{\succ} m$$

We shall focus on two particular *orders of preference* regarding the *result* Υ with respect to the numerical *results*.

Type P

If m and n are numerical *results* and $m < n$ then:

$$m \overset{P}{\succ} n \overset{P}{\succ} \Upsilon,$$

and consequently the opposite should hold for player E:

$$\Upsilon \overset{E}{\succ} n \overset{E}{\succ} m.$$

Here player P while wishing to minimize the numerical outcomes considers *nontermination* to be inferior to all other outcomes.
player E, on the other hand, while wishing to maximize the numerical outcomes considers *nontermination* to be superior to all other outcomes.

---

[11] The reader should note that in some problems it may not be appropriate to use a model that assigns fixed *roles* to the players. In the problem of *air combat*, for example, each party wants to obtain the opportunity to assume the role of the *pursuer*.

[12] A game of *pure conflict* was called by the founders of the classical game theory, Von Neumann and Morgenstern, a *zero sum game*.

Type E

If m and n are numerical *results* and $m < n$ then:

$$\Upsilon \overset{P}{\succ} m \overset{P}{\succ} n,$$

and consequently the opposite should hold for player E:

$$n \overset{E}{\succ} m \overset{E}{\succ} \Upsilon.$$

Here player P is still wishing to minimize the numerical outcomes but considers *nontermination* to be superior to all other outcomes.
player E, on the other hand, is also still wishing to maximize the numerical outcomes but considers *nontermination* to be inferior to all other outcomes[13].

2.7 Strategies

2.7.1 Decisions and Information

So far we discussed how do the players compare results of past decision sequences (*control histories*) as a part of a *posterior* analysis of a *play*. Obviously the players are more concerned with the question of how to make a proper decision in the course of an actual game rather than with the research of past events. The concept of *control history* does not suffice to satisfy this concern by itself since at an instance in the course of a game a *control history* is not yet uniquely determined! In order to proceed we need to investigate the nature of decision making in further depth.
Let us recall that all that a decision maker has at his disposal when he has to decide is *information*.
A *choice* of a control $\mathbf{u}$ from the set $\mathcal{U}$ of *admissible controls* can be considered as an *operation* D on the *information*:

$$\mathbf{u} = D(\imath_p),$$

where $\imath_p$ is the *state of information* of player P at the instance when the decision is made.
The vector $\imath_p$ is an element of a set $\mathcal{I}_P$ consisting of all feasible *states of information* in a game problem (See Figure 2.2). A rule (sometimes called *decision policy* or *control laws*) that assigns recommendations for choices (not necessarily unique!) of *control vectors* $\mathbf{u}$ from the set $\mathcal{U}$ of admissible controls to each element of a set of *states of information* is called a *strategy*

[13] Notice therefore that an order of preference over the numerical outcomes may be ambiguous unless an attitude toward nontermination is incorporated in it!

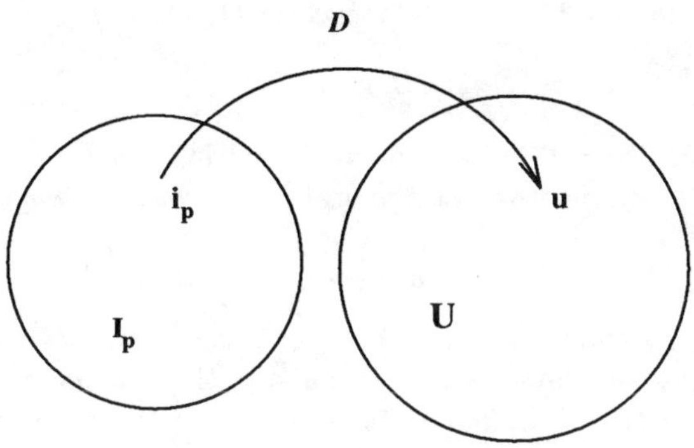

FIGURE 2.2. The map from information into control

for player P[14].
We shall use superscripted capital letters U (for example $\tilde{U}$) and V (for example $\tilde{V}$) to denote strategies of player P and player E respectively.

2.7.2 Realizations of Strategies

In order to qualify some strategies as *optimal strategies* we shall have to compare them with other strategies. A reasonable way to compare strategies is to compare the outcomes of plays in which the strategies were applied. We shall say that a *control history* $\tilde{\mathbf{u}}(\cdot)$ is a *realization* of the *strategy* $\tilde{U}$ in a play $[\mathbf{x}, \tilde{\mathbf{u}}(\cdot), \mathbf{v}(\cdot)]$ if at a state $\mathbf{x}(t)$ in the play the choice $\tilde{\mathbf{u}}(t)$ is recommended by the strategy $\tilde{U}$.

2.7.3 Strategies that Guarantee Nontermination

If player E considers the result Υ, (corresponding to *non-termination*) superior to any numerical outcome[15] he will certainly examine if there exist strategies that can guarantee him nontermination for games that start from all the states that belong to some subset of the game-set.
We can often verify that a candidate strategy can guarantee nontermination referring only to the nature of the *equations of motion*. Suppose, for

[14] The reader should be aware of the difference between *control histories* and *strategies*. A *control history* is associated with a course of a game only after it is played while a *strategy* may be available at any instance.

[15] We shall assume *order of preference of type P* throughout the book unless stated otherwise

the sake of simplicity, that the *target-set* is enclosed by a smooth surface[16]

$$B(\mathbf{x}) = 0.$$

Let $\mathbf{n}(\mathbf{x})$ be the normal to the surface B pointing away from the *target set*. If a *strategy* $\tilde{V}$ implies that at the state $\mathbf{x}$ player E always chooses $\tilde{\mathbf{v}}(\mathbf{x})$ and

$$\mathbf{f}[\mathbf{x}, \mathbf{u}, \tilde{\mathbf{v}}(\mathbf{x})] \cdot \mathbf{n}(\mathbf{x}) > 0 \quad \text{for all} \quad u \in \mathcal{U},$$

then player E, by choosing $\tilde{\mathbf{v}}(\mathbf{x})$ whenever $\mathbf{x}$ is on $B(\mathbf{x}) = 0$, can guarantee that trajectories of plays that start outside the set enclosed by $B(\mathbf{x}) = \mathbf{0}$ never cross $B(\mathbf{x}) = \mathbf{0}$ and thus never attain a *termination situation* on the *target-set*. The strategy $\tilde{V}$ is obviously a *startegy that guarantees nontermination* for all games that start outside the set enclosed by $B(\mathbf{x}) = \mathbf{0}$.

2.7.4 Strategies that Guarantee Termination

If player P regards the result Υ (that corresponds to *non-termination*) inferior to all the numerical results he will certainly examine if there exist strategies that can guarantee him termination for games that start from all the states that belong to some subset of the game set.
We can often verify that a candidate strategy can guarantee termination referring only to the nature of the *equations of motion*.
Consider a candidate strategy $\tilde{U}$ of player P that assigns the control $\tilde{\mathbf{u}}(\mathbf{x})$ whenever the state in a game is $\mathbf{x}$. One way to show that such a strategy $\tilde{U}$ is a strategy that guarantees termination for games that start in a certain region $\mathcal{W}$ in the game set is described in the following:

The s-function. Let $\mathbf{n}(\mathbf{x})$ be the normal to the boundary $\partial\mathcal{W}$ of the set $\mathcal{W}$ pointing into its exterior. Supppose that a nonnegative, bounded, single valued and smooth[17] function $s(\mathbf{x})$ that associates $s(\mathbf{x}) > 0$ to points $\mathbf{x} \in \mathcal{W}$ and $s(\mathbf{x}) = 0$ to points $\mathbf{x}$ that belong to the target set $\mathcal{C}$ is provided and that for some positive constant a the following is also true:

$$\mathbf{f}(\mathbf{x}, \tilde{\mathbf{u}}(\mathbf{x}), \mathbf{v}) \cdot \nabla s(\mathbf{x}) < -a < 0 \qquad \text{for all } \mathbf{v} \in \mathcal{V}, \tag{2.4}$$

and that for $\mathbf{x} \in \partial\mathcal{W}$ also:

$$\mathbf{f}(\mathbf{x}, \tilde{\mathbf{u}}(\mathbf{x}), \mathbf{v}) \cdot \mathbf{n}(\mathbf{x}) < 0 \qquad \text{for all } \mathbf{v} \in \mathcal{V}. \tag{2.5}$$

---

[16]The following argument can be applied with suitable modifications for much relaxed cases.

[17]This condition can be relaxed. The argument can be modified for piece-wise continuous and piece-wise smooth $s(\mathbf{x})$.

Discussion

We can now argue that:

- It follows from relation (2.5) that if a game starts at $\mathbf{x} \in \mathcal{W}$ the strategy $\tilde{U}$ guarantees player P that the state of the game will not leave the set $\mathcal{W}$.
- It follows from relation (2.4) that player P can guarantee, by realizing the strategy $\tilde{U}$, that $\dot{s}(\mathbf{x}) < -a < 0$ for all $\mathbf{x} \in \mathcal{W}$. This means that in all games that starts at $\mathbf{x} \in \mathcal{W}$ the function $s[\mathbf{x}(t)]$ decreases monotonously in a finite rate and penetrates $s(\mathbf{x}) = 0$ at a time $t_f(\mathbf{x}) < s(\mathbf{x})/a$. This is equivalent to the statement that the strategy $\tilde{U}$ guarantees that for a game that starts at $\mathbf{x} \in \mathcal{W}$ a definite terminating situation is attained in a finite time $t_f(\mathbf{x}) < s(\mathbf{x})/a$.
- Moreover, we argue that for any $\mathbf{x} \in \mathcal{W}$ if we are given any arbitrary function $\mathbf{v}(s)$ of admissible controls $\mathbf{v} \in \mathcal{V}$ for the domain $0 \leq s \leq s(\mathbf{x})$ the strategy $\tilde{U}$ assures that a terminating situation will be attained.

  We infer that all plays $[\mathbf{x}, \tilde{\mathbf{u}}(\cdot), \mathbf{v}(\cdot)]$ where $\tilde{u}(\cdot)$ is a realization of the strategy $\tilde{U}$ are *terminating plays*.

2.7.5 Admissibility of Strategies.

Not all the strategies are *admissible strategies*. A strategy is considered as a *nonadmissible strategy* if it causes a breach of a *state constraint* or a *control constraint* or if according to the rules of the game the information needed for its realization is not available to the player. Notice that a strategy can be a *nonadmissible strategy* even if it does not violate the *control constraints* $(\mathbf{u} \in \mathcal{U}\ ,\ \mathbf{v} \in \mathcal{V})$.

It is essential, from a theoretical as well as practical points of view, to define precisely which are admissible *states of information* (elements of the sets $\mathcal{I}_P$ and $\mathcal{I}_E$) and which are not.

We do that by specifying the so called *information constraints*:

The information constraints

Admissible states of information have to abide to the following assumptions:

Known rules

Both players have a *complete* knowledge of the rules of the game.

Perfect information about the state

Both players have *perfect information* about the current *state vector* and the current time at each instance of the game[18].

---

[18] Attempts were made by several author to formulate theories for differential games where the players may have partial or delayed information about the cur-

Causality

The players have only *causal information* (They do not have information that is related to the future).

Independence of controls

The players do not know their opponent's current decision[19].

A *strategy* is considered to be an *admissible strategy* if it satisfies the *information constraints*, the *control constraints* and the *state constraints*. Let us first examine two important classes of admissible strategies, *closed loop strategies* and *open loop strategies*[20].

Closed loop strategies

The class of *strategies* that assign the same recommendations for choices of *controls* to all the *information states* that share the same *state vector* is the class of *closed loop strategies.* In other words, in a *closed loop strategy* the rule for the choice of controls depends only the current state.

Open loop strategies

The class of *strategies* that assign the same recommendations for choices of *controls* to all the *information states* that share the same time (time may be measured from some common reference time) is the class of *open loop strategies.* In other words, in an *open loop strategy* the rule for the choice of controls depends only the current time[21].

Let us use the following simple example to demonstrate some admissible and nonadmissible strategies.

Example 2.7.1 (The Obstacle Tag game) *[Les85,Isb67]*

Player P pursues Player E in a plane that includes a circular obstacle with a radius of magnitude 1 into which neither player is allowed to enter. The velocity of the pursuer P is of magnitude $w > 1$ and the velocity of the evader E is of magnitude 1. Both have an unlimited rate of turn.
Player P wants to capture player E in a minimum time. We may formulate the game as follows: (see Figure 2.3).

The State Vector:

$$\mathbf{x} = (x, y, \theta).$$

---

rent state. Those issues are outside the scope of this book. We shall concentrate on *Two persons zero sum perfect information differential games.*

[19] This is one of the main differences between differential games and computer simulated war games.

[20] Those names originate from the theory of feedback control.

[21] It is assumed that the current time is known to the players as part of their information on the state.

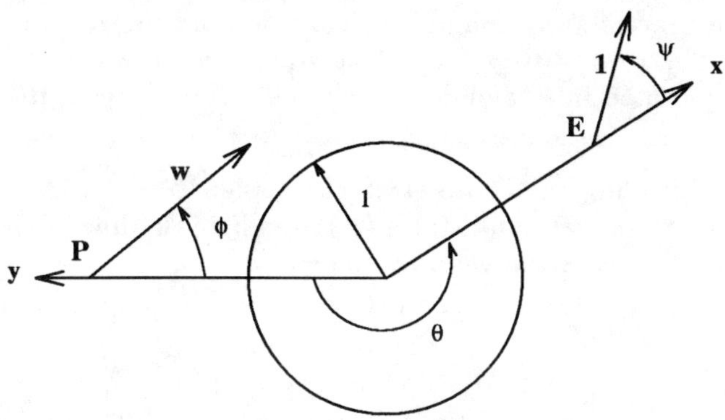

FIGURE 2.3. The Obstacle Tag game

The Game Set:

$$\mathcal{S} \equiv \{(x, y, \theta) \mid x \geq 1 \ , \ y \geq 1\}.$$

The Target Set:

$$\mathcal{C} \equiv \{(x, y, \theta) \mid x \geq 1 \ , \ y \geq 1 \ , \ x - y = 0 \ \}.$$

The Equations of Motion:

$$\dot{x} = -w \cos \phi,$$

$$\dot{y} = \cos \psi,$$

$$\dot{\theta} = \tfrac{\sin \psi}{y} + \tfrac{w \sin \phi}{x}.$$

Player P chooses ϕ and player E chooses ψ.

The Outcome Functional:

$$G(x, y, \theta) = 0 \quad , \ L(x, y, \theta, \phi, \psi) = 1$$

Case a

The following candidate strategies assign ϕ and ψ for states that belong to the set of all states $(x.y, \theta)$ where the *line of sight* between P and E crosses the *obstacle* and $\theta \neq \pi$.

$$\sin \phi = \begin{cases} -1/x & \text{if } \theta > \pi \\ 1/x & \text{if } \theta < \pi, \end{cases} \qquad |\phi| \leq \pi/2,$$

$$\sin \psi = \begin{cases} -1/y & \text{if } \theta > \pi \\ 1/y & \text{if } \theta < \pi, \end{cases} \qquad |\psi| \leq \pi/2$$

Both players are recommended to run along tangents to the obstacle. The rules satisfy the *information constraints* and each can be incorporated in a candidate *admissible closed loop strategy.*

Case b

The following candidate strategies assign ϕ and ψ for states that belong to the set of all states $(x.y,\theta)$ where the *line of sight* between P and E crosses the *obstacle* and $\theta = \pi$.

$$\sin\phi = -1/x \text{ or } 1/x, \qquad |\phi| \leq \pi/2$$

$$\sin\psi = \begin{cases} -1/y & \text{if } \sin\phi = -1/x\ . \\ 1/y & \text{if } \sin\phi = 1/x \end{cases}, |\psi| \leq \pi/2$$

Both players are recommended to run along tangents to the obstacle. Player P can choose between the upper tangent and the lower tangent but player E chooses to run along the upper tangent if player P chose the upper tangent and along the lower tangent if player P chose the lower.

The rule for player P satisfies the *information constraints* and can be incorporated in a candidate *admissible closed loop strategy* but the rule for player E, which depends on the choice made by player P, violates the *information constraints* and can not be part of an *admissible strategy.*

Case c

Let us consider a different candidate strategy for player E at states that belong to the set of all states $(x.y,\theta)$ where the *line of sight* between P and E crosses the *obstacle* and $\theta = \pi$.

Player E is recommended to choose a constant value $\tilde{\psi}$ for a duration δ regardless of the state during that period.

$$\sin\tilde{\psi} = \begin{cases} 1/y & \text{for } wy^2/x^2 > 1 \\ \\ wy/x^2 & \text{for } wy^2/x^2 \leq 1. \end{cases}$$

This is an *open loop* rule for a duration δ and it does not violate the *information constraints.*

•

2.8 Problems

Problem 2.8.1

Player P pursues player E in a plane. Both players have an unlimited rate of turn but their speeds are bounded: $v_P \leq A$, $v_E \leq B$.

1. Choose a *game-set* and a *state vector* representation.
2. What is the dimension of the problem?

•

Problem 2.8.2

Consider games with *fixed duration.*

1. Show that in a game with a fixed duration all the plays terminate.
2. what is the target-set?

•

Problem 2.8.3

Consider the problem of *Lion and Man.*

1. Show that the dimension of the problem of *lion and man* is 3.
2. Try to sketch the *game-set* and the *target-set.*

•

Problem 2.8.4

Consider a problem of *surveillance evasion.* A slow but more maneuverable boat tries to evade surveillance of a radar with a circular detection envelope that is mounted on a faster but less maneuverable ship. We shall assume that the boat is capable of instantaneous turns while the ship has a minimal turn radius.

1. Can we formulate such a problem as a differential game?
2. How does it correlate to the *Homicidal Chauffeur* problem?

•

Problem 2.8.5

Consider a problem of *collision avoidance.* Two ships try to avoid collision. Both have limited rates of turn.

1. Can we formulate this problem as a differential game? If not why?

•

Problem 2.8.6

One can argue that the knowledge of the current *state vector* together with the fact that the *trajectories* are assumed to be continuous imply that the players can compute the current decisions of their opponents (contradicting our information constraint!).

1. Think of an example that shows that this is not always true.

•

Problem 2.8.7

Consider the preview example of section 1.2.

1. Verify that if player E chooses $v = B$ then the choice $u = 3\pi/2$ provides player P with a better (lower) outcome than the choice of $\overset{*}{u}$. Use this example to realize the difference between the structure of the information upon which the decisions are made.

•

Problem 2.8.8

Equation (2.2) that defines a terminating situation, can be modified easily for the cases when $\mathbf{n}(\mathbf{x})$ has a jump discontinuity .

1. Perform the necessary modification.

•

Problem 2.8.9

Consider the case of a more generalized outcome functional,

$$\pi(x,t,u(\cdot),v(\cdot),G,T) =$$

$$G[x(T)]e^{-\int_t^T c(x,u,v,\tau)d\tau} + \int_t^T e^{-\int_t^\tau c(x,u,v,s)ds} L(x,u,v,\tau)d\tau.$$

1. Show that:

$$\pi(x,t,u(\cdot),v(\cdot),G,T) = \pi(x,s,u(\cdot),v(\cdot),G,T)e^{-\int_t^s c(x,u,v,\tau)d\tau} + \\ + \int_t^s e^{-\int_t^\tau c(x,u,v,z)dz} L(x,u,v,\tau)d\tau. \tag{2.6}$$

•

3

The Solution Concept

3.1 Introduction

The historic evidence about conflicts in real life shows that it is unwise to under-estimate an opponent. A cautious decision maker would like to base his decisions on an analysis that assumes a clever rival.

This idea is incorporated in the *solution concept* of differential games following the ideas of classical game theory regarding the notions of *state of information*, *strategy*, *equilibrium* and *value* for *zero sum games*.

Section 3.2 is devoted to the definitions of the *optimal strategies* and the *value function* of a differential game. The *optimal strategies* of player P are required to be *strategies that guarantee termination.* As we have mentioned in section 2.1 we try to refrain from putting undue restrictions on the mathematical properties of the control histories $\mathbf{u}(\cdot)$ and $\mathbf{v}(\cdot)$ as some writers do in order to provide for the existence of trajectories in the sense of the theory of ordinary differential equations. Instead, we prefer to impose stricter requirements on candidate *optimal strategies.*

Section 3.3 contains an extension of the notion of *optimal strategies* and a modification to the definition of the *value function* so that we may also consider sequences of *ϵ-optimal strategies* that satisfy a relaxed saddle inequality as candidate solutions for differential games.

Problems are given in section 3.4.

3.2 The Solution Quintet

The solution of a differential game is a quintet:

$$[\mathcal{W}, \mathcal{F}, \overset{*}{U}, \overset{*}{V}, J(\mathbf{x})].$$

$\mathcal{W}$ and $\mathcal{F}$ are subsets of the *game-set*,
$\overset{*}{U}$ is a strategy for player P,
$\overset{*}{V}$ is a strategy for player E,
$J(\mathbf{x})$ is a single valued real function defined on the set $\mathcal{W}$.

The properties of the *solution quintet* are:

Partitioning

The solution partitions the *game-set* to two mutually exclusive parts: The *escape-set* $\mathcal{F}$ and the *capture-set* $\mathcal{W}$. ($\mathcal{W}$ or $\mathcal{F}$ can be empty).

Optimal strategies

The strategy $\overset{*}{U}$ for player P is a *strategy that guarantees termination* for all the games that start at $\mathbf{x} \in \mathcal{W}$.

The strategy $\overset{*}{V}$ for player E is a *strategy that guarantees nontermination* for all the games that start at $\mathbf{x} \in \mathcal{F}$[1].

If a game starts at a state $\mathbf{x} \in \mathcal{W}$ and both players realize their optimal strategies the corresponding play, $[\mathbf{x}, \overset{*}{\mathbf{u}}(\cdot), \overset{*}{\mathbf{v}}(\cdot)]$, is called the *optimal play* or the *optimal trajectory* from $\mathbf{x}$ and the corresponding *outcome*, $\pi[\mathbf{x}, \overset{*}{\mathbf{u}}(\cdot), \overset{*}{\mathbf{v}}(\cdot)]$, is called the *optimal outcome* at $\mathbf{x}$.

Value

The function $J(\mathbf{x})$ is called the *value function* of the game.
It is required that the function $J(\mathbf{x})$ satisfy:

$$\begin{aligned} &J(\mathbf{x}) = \pi[\mathbf{x}, \overset{*}{\mathbf{u}}(\cdot), \overset{*}{\mathbf{v}}(\cdot)], \\ &\pi[\mathbf{x}, \overset{*}{\mathbf{u}}(\cdot), \mathbf{v}(\cdot)] \leq \pi[\mathbf{x}, \overset{*}{\mathbf{u}}(\cdot), \overset{*}{\mathbf{v}}(\cdot)] = J(\mathbf{x}) \overset{E}{\preceq} \pi[\mathbf{x}, \mathbf{u}(\cdot), \overset{*}{\mathbf{v}}(\cdot)], \end{aligned} \tag{3.1}$$

where
$[\mathbf{x}, \overset{*}{\mathbf{u}}(\cdot), \mathbf{v}(\cdot)]$ is any play with a realization of the strategy $\overset{*}{U}$
and $[\mathbf{x}, \mathbf{u}(\cdot), \overset{*}{\mathbf{v}}(\cdot)]$ is any play with a realization of the strategy $\overset{*}{V}$[2].
Relation (3.1) is called the *saddle inequality*.

A solution quintet is considered as an *admissible solution* if the *strategies* $\overset{*}{U}$ and $\overset{*}{V}$ are *admissible strategies*. Other wise we call it a *non-admissible solution*.

Let us discuss some properties of this *solution concept*.

Theorem 3.2.1 (Uniqueness of the value function)

If the *value function* of a differential game exists it is *unique*!

---

[1] Evidently it suffices that the strategy $\overset{*}{V}$ will be defined only on the boundaries of $\mathcal{F}$.

[2] We used $\overset{E}{\preceq}$ in relation (3.1) because a play $[\mathbf{x}, \mathbf{u}(\cdot), \overset{*}{\mathbf{v}}(\cdot)]$ may be a *nonterminating play* even for $\mathbf{x} \in \mathcal{W}$.

Proof

Let us assume the contrary.

Let $[\mathcal{W}, \mathcal{F}, \overset{*}{U}, \overset{*}{V}, J(\mathbf{x})]$ and $[\tilde{\mathcal{W}}, \tilde{\mathcal{F}}, \tilde{U}, \tilde{V}, \tilde{J}(\mathbf{x})]$ represent two different candidate solutions[3] where for some $\mathbf{x}$ we have $J(\mathbf{x}) \neq \tilde{J}(\mathbf{x})$.

Case a: $\mathbf{x} \in \mathcal{W} \cap \tilde{\mathcal{W}}$.

Both candidates satisfy the *saddle relations* (3.1) so we may write the following expressions:

$$\pi[\mathbf{x}, \overset{*}{\mathbf{u}}(\cdot), \tilde{\mathbf{v}}(\cdot)] \leq \pi[\mathbf{x}, \overset{*}{\mathbf{u}}(\cdot), \overset{*}{\mathbf{v}}(\cdot)] = J(\mathbf{x}) \overset{E}{\preceq} \pi[\mathbf{x}, \tilde{\mathbf{u}}(\cdot), \overset{*}{\mathbf{v}}(\cdot)], \tag{3.2}$$

$$\pi[\mathbf{x}, \tilde{\mathbf{u}}(\cdot), \overset{*}{\mathbf{v}}(\cdot)] \leq \pi[\mathbf{x}, \tilde{\mathbf{u}}(\cdot), \tilde{\mathbf{v}}(\cdot)] = \tilde{J}(\mathbf{x}) =\overset{E}{\preceq} \pi[\mathbf{x}, \overset{*}{\mathbf{u}}(\cdot), \tilde{\mathbf{v}}(\cdot)]. \tag{3.3}$$

Note that the left term in relation (3.2) equals the right term in relation (3.3), so evidently:

$$\tilde{J}(\mathbf{x}) = J(\mathbf{x})$$

Case b. $\mathbf{x} \in \mathcal{F} \cap \tilde{\mathcal{W}}$

From $\mathbf{x} \in \mathcal{F}$ it follows that the play $[\mathbf{x}, \tilde{\mathbf{u}}(\cdot), \overset{*}{\mathbf{v}}(\cdot)]$ is a *nonterminating* play, while from $\mathbf{x} \in \tilde{\mathcal{W}}$ it follows that *termination is guaranteed*.

$$\mathcal{F} \cap \tilde{\mathcal{W}} = \emptyset$$

follows.

•

Corollary 3.2.2

Let $[\tilde{\mathcal{W}}, \tilde{\mathcal{F}}, \tilde{U}, \tilde{V}, \tilde{J}(\mathbf{x})]$ be a candidate solution where $\tilde{U}$ or $\tilde{V}$ do not satisfy the *information constraints* for some of the information states. If an admissible solution exists then $\tilde{J}(\mathbf{x})$ is the *value function* of the game.

Proof

The proof follows immediately from the proof of Theorem 3.2.1.

•

The *saddle relation* (3.1) implies that a solution of a differential game has the property of *Nash equilibrium*. Neither player P nor player E can improve their guaranteed result, represented by the value $J(\mathbf{x})$, by a *unilateral deviation* from their optimal strategy. An interesting property of the *Nash equilibrium* is that a prior knowledge of the optimal strategy of the opponent is of no benefit[4].

---

[3] Not necessarily with admissible strategies!

[4] The reader should notice that sometimes a strategy provides several alternatives for the choice of controls at a certain information state so that a prior knowledge of a strategy does not imply prior knowledge of the choices of the controls!

We should notice that the *optimality* of $\overset{*}{\mathbf{u}}$ and $\overset{*}{\mathbf{v}}$ is in the sense of *guaranteed outcome*. By realizing an optimal strategy in a play that starts at a state $\mathbf{x}$ the player can guarantee an outcome which is at least as good as the value $J(\mathbf{x})$.
The strategies $\overset{*}{\mathbf{u}}$ or $\overset{*}{\mathbf{v}}$ might not be the best for taking advantage of a prior knowledge about a *deviation* from the optimal strategies by the opponent[5] (see Problem 2.8.7).

3.3 The Extended Solution Concept

In many cases we face the unfortunate situation that we can't find an *admissible solution* while we can construct a *candidate solution* $[\tilde{\mathcal{W}}, \tilde{\mathcal{F}}, \tilde{U}, \tilde{V}, \tilde{J}(\mathbf{x})]$ which must be considered a *nonadmissible solution* to the game since the optimal strategies $\tilde{U}$ or $\tilde{V}$ fail to satisfy the *information constraints* in a subset of the sets $\mathcal{W}$ or $\mathcal{F}$.
Recalling Corollary 3.2.2 we can consider $\tilde{J}(\mathbf{x})$ as the unique candidate for the value function $J(\mathbf{x})$ if a solution in admissible strategies does exist. We find ourselves stuck halfways. We know that if indeed a solution to the differential game exists then for the game that starts at the state $\mathbf{x}$ each player can guarantee himself an outcome $J(\mathbf{x})$ but we do not have admissible strategies that enable the players to realize this guaranteed outcome.

From the point of view of an application, an admissible strategy that can guarantee any player that the outcome will stay within a prespecified proximity of magnitude ϵ of the *ideal* guaranteed $J(\mathbf{x})$, is quite satisfactory in practice.

Those observations motivate us to consider an extension to the *regular solution concept* that will accommodate a notion of *ϵ-optimal strategies*.
We define the solution of a differential game in the sense of the *extended solution concept* as a quintet

$$[\mathcal{W}, \mathcal{F}, \{\overset{*}{U}_i\}, \{\overset{*}{V}_i\}, J(\mathbf{x})].$$

$\mathcal{W}$ and $\mathcal{F}$ are subsets of the *game-set*,
$\{\overset{*}{U}_i\}$ is a set of admissible strategies for player P,
$\{\overset{*}{V}_i\}$ is a set of admissible strategies for player E,
$J(\mathbf{x})$ is a single valued real function defined on the set $\mathcal{W}$.

The properties of the *solution quintet* are:

[5] Recall that we excluded such preknown information from our structure of the game model though espionage sometimes plays an important role in real conflicts.

Partitioning

The solution should partition the *game-set* to two mutually exclusive parts: the *escape-set* $\mathcal{F}$ and the *capture-set* $\mathcal{W}$. ($\mathcal{W}$ or $\mathcal{F}$ can be empty).

ϵ -optimal strategies

The strategies $\overset{*}{U}_i$ for player P are *strategies that guarantees termination* for all the games that start at $\mathbf{x} \in \mathcal{W}$.

The strategies $\overset{*}{V}_i$ for player E are *strategies that guarantees nontermination* for all the games that start at $\mathbf{x} \in \mathcal{F}$.

Value

The function $J(\mathbf{x})$ is called the *value function* of the game.
It is required that given any $\epsilon_P > 0$, there exists a positive number $n_P(\epsilon_P)$, such that for all numbers $i > n_P(\epsilon_P)$ the following has to hold for any state $\mathbf{x}$ in the *capture-set* $\mathcal{W}$:

$$\pi[\mathbf{x}, \overset{*}{\mathbf{u}}_i(\cdot), \mathbf{v}(\cdot)] \leq J(\mathbf{x}) + \epsilon_P, \tag{3.4}$$

and similarly, given any $\epsilon_E > 0$, there should exist a positive number $n_E(\epsilon_E)$, such that for all numbers $i > n_E(\epsilon_E)$ the following has to hold for any state $\mathbf{x}$:

$$J(\mathbf{x}) - \epsilon_E \overset{E}{\preceq} \pi[\mathbf{x}, \mathbf{u}(\cdot), \overset{*}{\mathbf{v}}_i(\cdot)], \tag{3.5}$$

where $[\mathbf{x}, \overset{*}{\mathbf{u}}_i(\cdot), \mathbf{v}(\cdot)]$ is any play with a realization of the strategy $\overset{*}{U}_i$ and $[\mathbf{x}, \mathbf{u}(\cdot), \overset{*}{\mathbf{v}}_i(\cdot)]$ is any play with a realization of the strategy $\overset{*}{V}_i$.

Any solution $[\mathcal{W}, \mathcal{F}, \overset{*}{U}, \overset{*}{V}, J(\mathbf{x})]$ in the sense of the *regular solution concept* of section 3.2 obviously qualifies as a solution in the sense of the *extended solution concept*.

We recall that in the case of the *regular solution concept* we have from Theorem 3.2.1 that if the value function $J(\mathbf{x})$ of a differential game exists it is unique. Next we shall show that this important property holds for the value function in the context of the *extended solution concept* as well.

Theorem 3.3.1 (Uniqueness of the extended value)

The *extended value function* $J(\mathbf{x})$ of a differential game, if it exists, is unique.

Proof

Let us again assume the contrary. Suppose that we have the two following candidate *solution quintets*:

$$[\mathcal{W}, \mathcal{F}, \{\overset{*}{U}_i\}, \{\overset{*}{V}_i\}, J(\mathbf{x})],$$

$$[\tilde{\mathcal{W}}, \tilde{\mathcal{F}}, \{\tilde{U}_i\}, \{\tilde{V}_i\}, \tilde{J}(\mathbf{x})].$$

Again we must have that $\tilde{\mathcal{W}} = \mathcal{W}$ and $\tilde{\mathcal{F}} = \mathcal{F}$ because otherwise there would exist a state $\mathbf{x}$ at which *termination* and *nontermination* are guaranteed at the same time ...

Assume that there exists a state $\mathbf{x}$ for which $\tilde{J}(\mathbf{x}) \neq J(\mathbf{x})$.

By specializing relation (3.4) and relation (3.5) we obtain for any given $\epsilon > 0$ that there exist $n(\epsilon)$ such that for all $i > n(\epsilon)$:

$$\tilde{J}(\mathbf{x}) - \epsilon \leq \pi[\mathbf{x}, \overset{*}{\mathbf{u}}_i(\cdot), \tilde{\mathbf{v}}_i(\cdot)] \leq J(\mathbf{x}) + \epsilon,$$

$$J(\mathbf{x}) - \epsilon \leq \pi[\mathbf{x}, \tilde{\mathbf{u}}_i(\cdot), \overset{*}{\mathbf{v}}_i(\cdot)] \leq \tilde{J}(\mathbf{x}) + \epsilon.$$

implying that for any $\epsilon > 0$:

$$\tilde{J}(\mathbf{x}) - 2\epsilon \leq J(\mathbf{x}) \leq \tilde{J}(\mathbf{x}) + 2\epsilon.$$

Since ϵ can be arbitrarily small it follows that $\tilde{J}(\mathbf{x}) = J(\mathbf{x})$.

•

The proof of the *Nash property* is left to the reader.

3.4 Problems

Problem 3.4.1 (The Berkovitch problem)

Consider the functional

$$\pi(u, v) = \frac{(u - v)^2}{2}, \quad |u| \leq 1, \quad |v| \leq 1.$$

player P chooses u and wants to minimize $\pi(u, v)$ while player E chooses v and wants to maximize $\pi(u, v)$.

1. Is the functional $\pi(u, v)$ separable?
2. Is this a differential game?
3. Show that $\min_u \{\max_v \pi(u, v)\} = 1/8$
4. Show that player E can guarantee an outcome of at least 1/8 only if $\overset{*}{v} = \overset{*}{v}(u)$ (Which means that player E has a prior knowledge of the choice of u made by player P).

5. Show that player P can guarantee an outcome of at most 0 only if $\overset{*}{u}=\overset{*}{u}(v)$ (Which means that player P has a prior knowledge of the choice of u made by player E).

•

Problem 3.4.2

Consider a differential game that takes place in the plane.

The game-set:

$$\mathcal{S} \equiv \{(x, y) \mid y > o\,,\; x + y - 1 \leq 0\,\}\,.$$

The target-set:

$$\mathcal{C} \equiv \{(x, y) \mid x + y = 1\,\}\,.$$

The equation of motion:

$$\dot{x} = 1 + v\,, \quad |v| \leq 1,$$

$$\dot{y} = 1 + u\,, \quad |u| \leq 1.$$

The outcome functional The game is a *terminal cost game* and

$$G(\mathbf{x}) = y^2 - x^2.$$

Player P, who chooses u, wants to minimize the outcome while player E, who chooses v, wants to maximize it.

Examine a play that starts at $(0, 0)$.

1. Show that if the order of preference is of type P the value is 1.
2. Show that if the order of preference is of type E the value is -1.

•

Problem 3.4.3 (The Wall Pursuit Game) [6]

An evader E, (*an escaping criminal* ...), is constrained to run along a straight wall. A pursuer, (*a policeman* ...) that has some speed advantage chases the evader. Capture occurs when P gets within a distance l from E.

Let us cast the problem into the format of a game model. We shall define the game in a moving *relative reference frame* where the x-axis lies along the wall and the y-axis passes through the real position of the pursuer.

---

[6] This problem was first described by R. Isaacs in his book (see [Isa75]).

The game-set:

$$\mathcal{S} \equiv \{(x,y) \mid y \geq o \ , \ \ x^2 + y^2 \geq l^2\}.$$

The target-set:

$$\mathcal{C} \equiv \{(x,y) \mid y \geq o \ , \ \ x^2 + y^2 = l^2\}.$$

The equations of motion:

$$\dot{x} = \psi - w \cos\varphi \ , \quad |\psi| \leq 0,$$

$$\dot{y} = -w \sin\varphi,$$

$w > 1$ is a constant. Player P chooses φ and player E chooses ψ.

The outcome functional:

$$G(x,y) = 0, \quad L(x,y,\varphi,\psi) = 1. \tag{3.6}$$

This is a *pure pursuit evasion game.*

Consider the case when $l = 0$ (the case of *point capture*). It is well known that it is optimal for player P to steer always toward the *collision point.* On an optimal trajectory both players use constant controls. The value function $J(x,y)$ is given by

$$J(x,y) = \tau(x,y),$$

where $\tau(x,y)$ is the smallest positive solution of[7]

$$(x + \tau)^2 + y^2 = (w\tau)^2. \tag{3.7}$$

1. Argue why for games that start at $(0,y)$ player P do not have an admissible strategy that can realize the outcome $J(0,y)$.

2. Try to suggest ϵ-optimal strategies for player P.

[7] Consider the triangle defined by the initial locations of the players and the collision point in the real plane of pursuit. Equation (3.7) can be obtained from this triangle by inspection.

4

Semipermeability of Surfaces

4.1 Introduction

This chapter is devoted to a preliminary discussion of *semipermeable surfaces*. (We shall abbreviate this long name by the notation *SPS*). *R. Isaacs* was the first to note the extreme relevance of the property of semipermeability to the theory of differential games and devoted much of his monumental work [Isa75] to it. He realized that a surface in the game-set of a *pursuit evasion game* that separates a *capture zone* from an *escape zone* must have the property that each player should be able to prevent the other player from crossing it in the unwanted sense[1]. Another important contribution was made by *P. Bernhard* who was first to point out the necessity to impose extra conditions in order to provide for semipermeability at nonsmooth corners of *composite semipermeable surfaces*.
In this chapter we give the classical definitions and an illustrative example for the smooth case. We use the same example to demonstrate a case with a *leaking corner*. Necessary conditions for a non leaking semipermeable corner are discussed.

4.2 Smooth Semipermeable Surfaces

Let $\mathbf{n}(\mathbf{x})$ be the *normal* to a smooth $(n-1)$-dimensional surface $D(\mathbf{x}) = 0$ at the point $\mathbf{x}$. The point $\mathbf{x}$ can be approached from the two sides of the surface $D(\mathbf{x})$. We shall denote the side that corresponds to the positive sense of the normal vector $\mathbf{n}(\mathbf{x})$ as *side-2* and the side that corresponds to the negative sense of $\mathbf{n}(\mathbf{x})$ will be called *side-1*.

We shall say that the surface D is crossed at the point $\mathbf{x}$ by a velocity vector $\mathbf{f}(\mathbf{x},\mathbf{u},\mathbf{v})$ from its *positive* side-2 to its *negative* side-1 if we have

$$\mathbf{f}(\mathbf{x},\mathbf{u},\mathbf{v}) \cdot \mathbf{n}(\mathbf{x}) < 0,$$

and similarly, we shall say that the surface D is crossed at the point $\mathbf{x}$ by a velocity vector $\mathbf{f}(\mathbf{x},\mathbf{u},\mathbf{v})$ from its *negative* side-1 to its *positive* side-2 if

[1]The idea of investigating the behavior and the stability of a system by analyzing its velocity vector with respect to the state space is not particular to differential games and is even not new, (recall the approaches of *Lyapunov* and *Bendixon-Poincare* in the theory of ordinary differential equations).

we have

$$\mathbf{f}(\mathbf{x},\mathbf{u},\mathbf{v})\cdot\mathbf{n}(\mathbf{x}) > 0.$$

We assume, with no loss of generality, that player E prefers side-2 of the surface D while player P prefers to stay in side-1.

Let us consider the case that at $\mathbf{x}\in D$ there exists a control $\overset{*}{\mathbf{v}}\in\mathcal{V}$ such that:

$$\mathbf{f}(\mathbf{x},\mathbf{u},\overset{*}{\mathbf{v}})\cdot\mathbf{n}(\mathbf{x}) \geq \mathbf{0} \quad , \quad \text{for all } \ \mathbf{u}\in\mathcal{U}. \tag{4.1}$$

Equation (4.1) means that player E can prevent the crossing of the surface D at $\mathbf{x}$ from side-2 to side-1.

Similarly, if there exists a control $\overset{*}{\mathbf{u}}\in\mathcal{U}$ such that:

$$\mathbf{f}(\mathbf{x},\overset{*}{\mathbf{u}},\mathbf{v})\cdot\mathbf{n}(\mathbf{x}) \leq \mathbf{0} \quad , \quad \text{for all } \ \mathbf{v}\in\mathcal{V}.$$

We shall say that player P can prevent the crossing of the surface D at $\mathbf{x}$ from side-1 to side-2.

If both such $\overset{*}{\mathbf{u}}$ and $\overset{*}{\mathbf{v}}$ exist then:

$$\begin{array}{lll}\mathbf{f}(\mathbf{x},\overset{*}{\mathbf{u}},\mathbf{v})\cdot\mathbf{n}(\mathbf{x}) \leq & \mathbf{f}(\mathbf{x},\overset{*}{\mathbf{u}},\overset{*}{\mathbf{v}})\cdot\mathbf{n}(\mathbf{x}) = 0 \ \leq & \mathbf{f}(\mathbf{x},\mathbf{u},\overset{*}{\mathbf{v}})\cdot\mathbf{n}(\mathbf{x}), \\ \forall\mathbf{v}\in\mathcal{V} & & \forall\mathbf{u}\in\mathcal{U}\end{array} \tag{4.2}$$

which means that at $\mathbf{x}$ player P can prevent the crossing of the surface D from the side-1 to side-2 and at the same time player E can prevent the crossing of D from side-2 to side-1. We shall then say that the surface D is *semipermeable* at $\mathbf{x}$.

If $\overset{*}{\mathbf{u}}$ and $\overset{*}{\mathbf{v}}$ exist (not necessarily unique) such that the relation (4.2) holds for all $\mathbf{x}\in D$ we shall say that the surface D is a *semipermeable surface (SPS)* and that $(\overset{*}{\mathbf{u}},\overset{*}{\mathbf{v}})$ is the corresponding pair of *semipermeable controls* (denoted *SP-controls* and also called *barrier controls*).

The functions $\overset{*}{\mathbf{u}}$ $(\mathbf{x})$ and $\overset{*}{\mathbf{v}}$ $(\mathbf{x})$ may be regarded as *virtual closed loop strategies*. The trajectories of the ordinary differential equation

$$\dot{x} = \mathbf{f}(\mathbf{x},\overset{*}{\mathbf{u}},\overset{*}{\mathbf{v}}) \quad , \quad \mathbf{x}(0)\in D,$$

generate a path that lies on the semipermeable surface and do not leave it. Actually, the surface D is a *manifold* of such paths.

It is remarkable that the property of semipermeability does not depend on the *target-set* or on the *outcome functional*. If a semipermeable surface is identified in the state space it is common to all the differential games that have the same *equations of motion* regardless of the *target-set* and the *outcome functional*!

Example 4.2.1 (The Servo problem) *(see P. Bernhard in [Ber77])*

Consider the following differential game:

The game-set:

$$\mathcal{S} \equiv \{(x,y) \mid |y| \leq 1\}.$$

The target-set:

$$\mathcal{C} \equiv \mathcal{C}^{(+)} \cup \mathcal{C}^{(-)},$$

Where:

$$\mathcal{C}^{(+)} \equiv \{(x,y) \mid y = 1\},$$

$$\mathcal{C}^{(-)} \equiv \{(x,y) \mid y = -1\}.$$

The Equations of Motion:

$$\dot{x} = v/2q\ , \quad |v| \leq 1,$$

$$\dot{y} = u - x\ , \quad |u| \leq 1.$$

Player P chooses u and player E chooses v. $q > 0$ is a parameter. Let $\frac{1}{2} \leq q \leq 1$.

The Outcome Functional:

$$L(x,y,u,v) = 1, \quad G(x,y) = 0$$

Consider the surfaces D_1 and D_2

$$D_1(x,y) = -(1+x)^2 + (1+y)/q = 0,$$

$$D_2(x,y) = -(1-x)^2 + (1-y)/q = 0,$$

and their *normals* $\mathbf{n}_1(x,y)$ and $\mathbf{n}_2(x,y)$

$$\mathbf{n}_1(x,y) = -(1+x)\hat{\mathbf{e}}_x + \tfrac{1}{2q}\hat{\mathbf{e}}_y,$$

$$\mathbf{n}_2(x,y) = (1-x)\hat{\mathbf{e}}_x - \tfrac{1}{2q}\hat{\mathbf{e}}_y,$$

where $\hat{\mathbf{e}}_x$ and $\hat{\mathbf{e}}_y$ are unit vectors in the directions of x and y.

The reader can verify easily that the pair of *SP-controls*

$$\overset{*}{u}(x,y) = 1, \quad \overset{*}{v}(x,y) = 1,$$

provide that the branch $y \leq 1$ of $D_2(x,y)$ is semipermeable.
Similarly, the pair of *SP-controls*

$$\overset{*}{u}(x,y) = -1, \quad \overset{*}{v}(x,y) = -1,$$

provide that the branch $y \geq -1$ of $D_1(x,y)$ is semipermeable (see Figure 4.1).

•

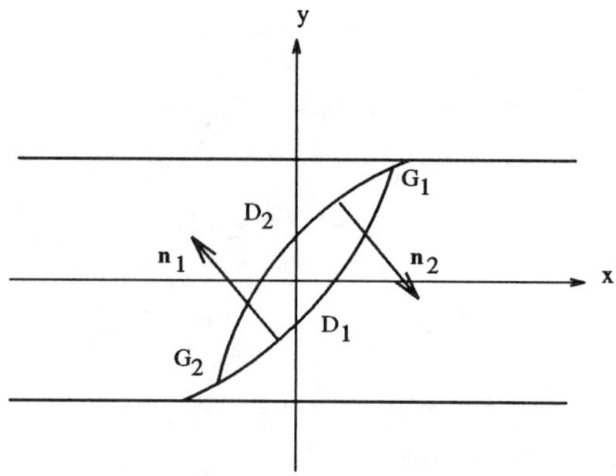

FIGURE 4.1. Semipermeable surfaces in the servo problem

4.3 Semipermeability of Composite Surfaces

4.3.1 LEAKING CORNERS

Let a *composite* surface $D(\mathbf{x}) = 0$ consist of two smooth $(n-1)$-dimensional semipermeable surfaces $D_1(\mathbf{x})$ and $D_2(\mathbf{x})$ that intersect along a $(n-2)$-dimensional corner surface G. Let $\mathbf{n}_1(\mathbf{x})$ and $\mathbf{n}_2(\mathbf{x})$ be the normals to D_1 and D_2 respectively. The composite surface D is in general non-smooth along the corner surface G where the normal function may have a *jump discontinuity*.

If, at the point $\mathbf{x} \in G$, player E wants to prevent the crossing from side-2 to side-1 of both $D_1(\mathbf{x})$ and $D_2(\mathbf{x})$ he should have a control $\overset{*}{\mathbf{v}}(\mathbf{x})$ such that:

$$\begin{aligned} \mathbf{f}(\mathbf{x},\mathbf{u},\overset{*}{\mathbf{v}}) \cdot \mathbf{n}_1(\mathbf{x}) \geq 0, \quad \forall \mathbf{u} \in \mathcal{U}, \\ \mathbf{f}(\mathbf{x},\mathbf{u},\overset{*}{\mathbf{v}}) \cdot \mathbf{n}_2(\mathbf{x}) \geq 0, \quad \forall \mathbf{u} \in \mathcal{U}. \end{aligned} \tag{4.3}$$

We can gain more geometric insight if we reformulate relation (4.3) as:

$$\mathbf{f}(\mathbf{x},\mathbf{u},\overset{*}{\mathbf{v}}) \cdot [\epsilon\mathbf{n}_1(\mathbf{x}) + (1-\epsilon)\mathbf{n}_2(\mathbf{x})] \geq 0, \quad \forall \mathbf{u} \in \mathcal{U},$$

where $0 \leq \epsilon \leq 1$.

Similarly, if at the point $\mathbf{x} \in G$, player P wants to prevent the crossing from side-1 to side-2 of both $D_1(\mathbf{x})$ and $D_2(\mathbf{x})$ he should have a control $\overset{*}{\mathbf{u}}(\mathbf{x})$ such that:

$$\begin{aligned} \mathbf{f}(\mathbf{x},\overset{*}{\mathbf{u}},\mathbf{v}) \cdot \mathbf{n}_1(\mathbf{x}) \leq 0, \quad \forall \mathbf{v} \in \mathcal{V}, \\ \mathbf{f}(\mathbf{x},\overset{*}{\mathbf{u}},\mathbf{v}) \cdot \mathbf{n}_2(\mathbf{x}) \leq 0, \quad \forall \mathbf{v} \in \mathcal{V}, \end{aligned} \tag{4.4}$$

or equivalently:

$$\mathbf{f}(\mathbf{x}, \overset{*}{\mathbf{u}}, \mathbf{v}) \cdot [\epsilon \mathbf{n_1}(\mathbf{x}) + (1-\epsilon)\mathbf{n_2}(\mathbf{x})] \leq 0, \quad \forall \mathbf{v} \in \mathcal{V},$$

where $0 \leq \epsilon \leq 1$.

If such $\overset{*}{\mathbf{u}}$ and $\overset{*}{\mathbf{v}}$ are not available to both players then we say that the corner *leaks*.

Example 4.3.1 (The Servo problem)
Let us reconsider Example 4.2.1.

The semipermeable surfaces $D_1(x,y)$ and $D_2(x,y)$ form two corners, at the point G_1 and G_2. (see Figure 4.1).
Let us check if player E can prevent the crossing of both $D_1(x,y)$ and $D_2(x,y)$ at the point G_1. It is easy to show that this requires:

$$\overset{*}{v} \geq 1 \quad \text{as well as} \quad \overset{*}{v} \leq \tfrac{1-x}{1+x}.$$

Recalling that at the corner G_1 we have $0 < x < 1$ we can conclude that those requirements cannot be satisfied simultaneously and that the corner *leaks* at G_1!

•

4.3.2 A Modified Definition of Semipermeability

We can use the idea of the theory of *viscosity solutions* to the *Isaacs PDE* (It will be discussed again in section 5.5.4) to state a definition of *semipermeability* in a *viscosity form* that will hold for both smooth and nonsmooth semipermeable surfaces.

Let $\mathcal{H}(\mathbf{x})$ denote any neighborhood of $\mathbf{x}$ and let $W(\mathbf{x})$ be any C^1 function defined in $\mathcal{H}(\mathbf{x})$. The function $Z(\mathbf{x})$ is defined in $\mathcal{H}(\mathbf{x})$ as:

$$Z(\mathbf{x}) = D(\mathbf{x}) - W(\mathbf{x}).$$

The surface $D(\mathbf{x})$ is semipermeable if we have *SP-controls* $(\overset{*}{\mathbf{u}}, \overset{*}{\mathbf{v}})$ such that for any $\mathbf{x} \in D$:

1. For any C^1 function $W(\mathbf{x})$ such that the function $Z(\mathbf{x})$ has a local maximum[2]

$$\begin{array}{l} \mathbf{f}(\mathbf{x}, \mathbf{u}, \overset{*}{\mathbf{v}}) \cdot \nabla \mathbf{W}(\mathbf{x}) \geq 0. \\ \forall \mathbf{u} \in \mathcal{U} \end{array} \tag{4.5}$$

[2]Assuming that the sense of $\nabla W(\mathbf{x})$ is aligned with the side of $D(\mathbf{x})$ that is preferred by player E.

2. For any C^1 function $W(\mathbf{x})$ such that the function $Z(\mathbf{x})$ has a local minimum at $\mathbf{x}$ we have that

$$\begin{array}{l}\mathbf{f}(\mathbf{x}, \overset{*}{\mathbf{u}}, \mathbf{v}) \cdot \nabla \mathbf{W}(\mathbf{x}) \leq \mathbf{0}. \\ \forall \mathbf{v} \in \mathcal{V}\end{array} \tag{4.6}$$

4.4 Problems

Problem 4.4.1 (The Bang-Bang-Bang Problem)

Consider the following problem:

The Game Set:

$$\mathcal{S} \equiv \{(x, y) \mid y \geq -1\}.$$

The Terget Set:

$$\mathcal{C} \equiv \{(x, y) \mid y = -1\}.$$

The Equations of Motion:

$$\dot{x} = -\varphi\tfrac{1}{2}\sqrt{3}y + \sqrt{3}(2 - \tfrac{1}{2}y) + 2\sqrt{3}\cos\psi, \qquad |\varphi| \leq 1,$$

$$\dot{y} = \varphi(3 + \tfrac{1}{2}y) + (\tfrac{1}{2}y - 1) + 2\sqrt{3}\sin\psi.$$

The Outcome Functional:
The payoff is terminal:

$$G(x, y) = -x \ , \quad L(x, y) = 0.$$

Player P wants to terminate and wishes to minimize the outcome and player E wishes to maximize the outcome. Player P chooses φ and player E chooses ψ.

1. Show that the lines of the family $x = c$ are semipermeable for $y \geq 0$.
2. Show that the lines of the family $y = -x/\sqrt{3} + c$ are semipermeable for $y < 1$.
3. Investigate the nature of the corners between lines from the two families.

•

5

Necessary Conditions

5.1 Introduction

In chapter 3 we introduced *solution concepts* of differential games. Some of those concepts like the *saddle inequality*

$$\pi[\mathbf{x}, \overset{*}{\mathbf{u}}(\cdot), \mathbf{v}(\cdot)] \leq \pi[\mathbf{x}, \overset{*}{\mathbf{u}}(\cdot), \overset{*}{\mathbf{v}}(\cdot)] = J(\mathbf{x}) \overset{E}{\preceq} \pi[\mathbf{x}, \mathbf{u}(\cdot), \overset{*}{\mathbf{v}}(\cdot)] \tag{5.1}$$

depend on *global* terms like $[\mathbf{x}, \overset{*}{\mathbf{u}}(\cdot), \mathbf{v}(\cdot)]$ (any play with a realization of the strategy $\overset{*}{U}$) or $[\mathbf{x}, \mathbf{u}(\cdot), \overset{*}{\mathbf{v}}(\cdot)]$ (any play with a realization of the strategy $\overset{*}{V}$), that refer to *complete* plays and are not easy to utilize for constructing candidate solutions.

In this chapter we shall induce *local properties* of the solutions of differential games from those global concepts. Those properties constitute *necessary conditions* to be satisfied by any *candidate solution*.
The necessary conditions discussed in this chapter suffice for devicing a method (to be discussed in chapter 7) for the *construction* of candidate solutions that do not involve discontinuities in $J(\mathbf{x})$ or in $\nabla J(\mathbf{x})$[1].
Let us preview the contents of this chapter.
The first section, section 5.2, deals with the properties of the *target-set*.
Subsection 5.2.1 defines the properties of the part of the *target-set*, (named by Isaacs as *the usable part*) , where optimal plays can terminate.
We proceed, in subsection 5.2.2, to show the relations that must hold on the part of *usable part* where optimal plays terminate between the *value function* $J(\mathbf{x})$ and the *terminal cost function* $G(\mathbf{x})$.
In section 5.3 we show that the surface that separates the *capture-set* from the *escape-set* should be a *nonleaking semipermeable surface*. The argumentation for this property holds for a much wider definition of differential games than the classical one.
Section 5.4 describes the properties of *optimal trajectories*.
The *principle of optimality* is the subject of subsection 5.4.1. Intuitively we feel that any segment of an optimal trajectory should be optimal regardless of the specific play in which it is incorporated. It seems obvious but it has to be proved. We give the proof with respect to the classical definition of

---

[1]The case when there are discontinuities in $J(\mathbf{x})$ or in $\nabla J(\mathbf{x})$ is covered in chapter 10.

the *outcome functional* but the property holds for more general outcome functionals.
A remarkable property of the *value function*, the fact that along optimal trajectories $J(\mathbf{x})$ is continuous and that $\frac{d}{dt}J(\mathbf{x})$ exists, is investigated in subsections 5.4.2 and 5.4.3.
In subsection 5.4.4 we introduce the *Hamiltonian* function.
Section 5.5 is dedicated to the *Isaacs equations*.
The subject of subsection 5.5.1 is probably the most important property of solutions of differential games. We compare plays where one play is an *optimal play* and the other is a play with a *semilocal deviation* from optimality.
Based on this property we derive the famous *Isaacs' equations* (subsection 5.5.2), discuss their *hodograph representation* (subsection 5.5.3) and give their more novel form, *the viscosity conditions* (subsection 5.5.4).
The adjoint equations are discussed in detail in section 5.6.
We end the chapter with some problems.

5.2 Properties of the Target Set

5.2.1 Partitioning of the Target Set

At points of the target-set where optimal trajectories terminate it is necessary to have the property that player P has a choice of control that guaurantees a *terminating situation* regardless of the choice of control by player E. In view of this property we can partition the target-set to three disjoint parts:

The usable part[2]

The *usable part* (the *UP*) is the part of the target-set where player P can guarantee termination regardless of the chöice of the control $\mathbf{v}$ by player E.

$$\mathrm{UP} \equiv \left\{\mathbf{x} \mid \mathbf{x} \in \mathcal{C},\ \exists\, \overset{*}{\mathbf{u}} \in \mathcal{U} \text{ with } \max_{\mathbf{v}\in\mathcal{V}} \left(\mathbf{f}(\mathbf{x}, \overset{*}{\mathbf{u}}, \mathbf{v}) \cdot \mathbf{n}(\mathbf{x})\right) < 0\right\}.$$

The nonusable part

The *nonusable part* (the *NUP*) is the part of the target-set where player E can guarantee nontermination regardless of the choice of the control $\mathbf{u}$ by player P.

$$\mathrm{NUP} \equiv \left\{\mathbf{x} \mid \mathbf{x} \in \mathcal{C},\ \exists\, \overset{*}{\mathbf{v}} \in \mathcal{V} \text{ with } \min_{\mathbf{u}\in\mathcal{U}} \left[\mathbf{f}(\mathbf{x}, \mathbf{u}, \overset{*}{\mathbf{v}}) \cdot \mathbf{n}(\mathbf{x})\right] > 0\right\}.$$

The boundary of the usable part

The *boundary of the usable part* (the *BUP*) is the part of the target-set where

$$\max_{\mathbf{v} \in \mathcal{V}} \left\{ \mathbf{f}(\mathbf{x}, \overset{*}{\mathbf{u}}, \mathbf{v}) \cdot \mathbf{n}(\mathbf{x}) \right\} = \quad 0 \quad = \quad \min_{\mathbf{u} \in \mathcal{U}} \left\{ \mathbf{f}(\mathbf{x}, \mathbf{u}, \overset{*}{\mathbf{v}}) \cdot \mathbf{n}(\mathbf{x}) \right\}.$$

In the case of *pure pursuit evasion games*, where $L(\mathbf{x}, \mathbf{u}, \mathbf{v}) = 1$ and $G(\mathbf{x}) = 0$, we have the following necessary condition:

Lemma 5.2.1

Optimal trajectories do not terminate on the NUP.

Proof
In the case of *pure pursuit evasion games* player E can obviously get a better result by delaying termination as much as he can. On the NUP player E has the capability to delay termination and it would be nonoptimal if he does not use it.

•

Example 5.2.1 (The Dolichobrachistochrone problem) [3]

The game-set:
The game takes place in a quadrant of a plane.

$$\mathcal{S} \equiv \{(x, y) \mid x \geq 0, \ y \geq o\}.$$

The target-set:

$$\mathcal{C} \equiv \{(x, y) \mid x = 0, \ y \geq o\}.$$

The equations of motion:

$$\dot{x} = \sqrt{y} \cos \varphi + \tfrac{1}{2} w(\psi + 1), \quad |\psi| \leq 1,$$

$$\dot{y} = \sqrt{y} \sin \varphi + \tfrac{1}{2} w(\psi - 1).$$

Player P chooses φ and player E chooses ψ.

The outcome functional:
The game is a pure pursuit evasion game:

$$L(x, y, \varphi, \psi) = 1, \ \ G(x, y) = 0.$$

---

[3] The problem was first stated by *Isaacs* [Isa75] but its complete solution was given (but not published) by *J. V. Breakwell* and published by *S. A. Chigir* [Chi76].

Let us partition the *target-set*. It is easy to see that the choice of $\tilde{\psi} = 1$ guarantees player E that no play will terminate at points on the target-set where $y < w^2$:

$$\text{NUP} \equiv \left\{(x, y) \mid x = 0,\ y < w^2\right\},$$

$$\text{UP} \equiv \left\{(x, y) \mid x = 0,\ y > w^2\right\},$$

$$\text{BUP} \equiv \left\{(x, y) \mid x = 0,\ y = w^2\right\}.$$

•

5.2.2 The Relation Between the Value and the Terminal Cost Functions

Not all the points of the *usable part (UP)* of the *target-set* are end points of optimal trajectories. We have reliable information about this only after a *complete* solution is constructed. However we have the following necessary condition concerning those points of the *UP* where optimal trajectories do terminate:

Lemma 5.2.2

At points $\mathbf{x}$ on the *usable part* where optimal trajectories terminate we have:

$$J(\mathbf{x}) = G(\mathbf{x}).$$

Proof

Let us recall that the definition of the outcome functional (see section 2.6) is:

$$\pi[\mathbf{x}_0, \mathbf{u}(\cdot), \mathbf{v}(\cdot)] = \int_0^{t_f} L[\mathbf{x}(\tau), \mathbf{u}(\tau), \mathbf{v}(\tau)]\, d\tau + G[\mathbf{x}(t_f)],$$

and that in section 2.3 we assumed that the ranges of the components of $\mathbf{f}(\mathbf{x}, \mathbf{u}, \mathbf{v})$ and of $L(\mathbf{x}, \mathbf{u}, \mathbf{v})$ are bounded over the domains of their independent variables.

The pair $[\overset{*}{\mathbf{u}}(t_f), \overset{*}{\mathbf{v}}(t_f)]$ provides a terminating situation at $\mathbf{x}(t_f)$ so:

$$\pi[\mathbf{x}(t_f), \overset{*}{\mathbf{u}}(t_f), \overset{*}{\mathbf{v}}(t_f)] = G[\mathbf{x}(t_f)].$$

•

Corollary 5.2.3

If the subset of the *usable part* where optimal trajectories terminate is of dimension m and if $\mathbf{t}_i$ are m vectors that span the tangent to that subset at $\mathbf{x}$ then the following m relations between the directional derivatives of $J(\mathbf{x})$ and $G(\mathbf{x})$ hold:

$$\nabla J(\mathbf{x}) \cdot \mathbf{t}_i = \nabla G(\mathbf{x}) \cdot \mathbf{t}_i \tag{5.2}$$

•

5.3 Semipermeability of the Boundary of the Escape Set $\mathcal{F}$

In section 3.2 we saw that a solution of a differential game induces a partitioning of the game-set to two disjoint and mutually exclusive parts: *capture-set* $\mathcal{W}$ and *escape-set* $\mathcal{F}$. The boundary that separates the sets $\mathcal{W}$ and $\mathcal{F}$ is obviously a *barrier surface* (a surface across which the *value function*, $J(\mathbf{x})$, is discontinuous).
A point $\mathbf{x}$ at which player E can apply a *strategy that guarantees nontermination* must belong to the *escape-set* $\mathcal{F}$ and points $\mathbf{x}$ at which player P can apply a *strategy that guarantees termination* must belong to the capture-set $\mathcal{W}$.
Naturally we should have the property that an optimal trajectory that starts in the capture-set $\mathcal{W}$ will never cross to the escape-set $\mathcal{F}$ and that an optimal play by player E does not allow any trajectory that starts inside the escape-set to cross into the capture-set.
This property is equivalent to the following lemma:

Lemma 5.3.1

The boundary $\partial\mathcal{F}$ of the *escape-set* $\mathcal{F}$ is a *semipermeable surface*.

Proof
Let $\mathbf{x}$ be a point on the surface $\partial\mathcal{F}$ and let $\mathbf{n}(\mathbf{x})$ be the normal to the surface $\partial\mathcal{F}$ at the point $\mathbf{x}$ pointing into the interior of the *escape-set* $\mathcal{F}$.
Suppose that the lemma is not true and that at some $\mathbf{x}$ on $\partial\mathcal{F}$ there exists a control $\tilde{\mathbf{v}} \in \mathcal{V}$ such that:

$$\mathbf{f}(\mathbf{x}, \overset{*}{\mathbf{u}}, \tilde{\mathbf{v}}) \cdot \mathbf{n}(\mathbf{x}) > 0.$$

Then there exists a nonterminating play, $[\mathbf{x}^{(-)}, \overset{*}{\mathbf{u}}(\cdot), \mathbf{v}(\cdot)]$, that crosses from $\mathbf{x}^{(-)}$ which is in the *capture-set* $\mathcal{W}$ to $\mathbf{x}^{(+)}$ which is in the *escape-set* $\mathcal{F}$. Evidently:

$$\pi[\mathbf{x}^{(-)}, \overset{*}{\mathbf{u}}(\cdot), \mathbf{v}(\cdot)] = \Upsilon > J(\mathbf{x}^{(-)}),$$

contradicting the saddle property of the solution.
We can show that assuming that at some $\mathbf{x} \in \partial\mathcal{F}$ there exists a control $\tilde{\mathbf{u}} \in \mathcal{U}$ such that:

$$\mathbf{f}(\mathbf{x}, \tilde{\mathbf{u}}, \overset{*}{\mathbf{v}}) \cdot \mathbf{n}(\mathbf{x}) < 0,$$

leads to a similar contradiction.

•

Remark:
Sometimes part of the *NUP* (the nonusable part of the target-set) belong to the *escape-set*. In such cases the corners between the *NUP* and the separating surface should not leak!

•

Corollary 5.3.2

An optimal trajectory that starts inside the *capture-set* $\mathcal{W}$ remains there till its termination. (Or equivalently: the boundary of the *escape-set* $\mathcal{F}$ constitutes a *closed barrier*).

•

5.4 Properties of Optimal Trajectories

5.4.1 Principle of Optimality (weak)

Let an optimal trajectory $[\mathbf{x}, \overset{*}{\mathbf{u}}, \overset{*}{\mathbf{v}}]$ start from the state $\mathbf{x}$ at the time t and terminate on the target-set at the state $\mathbf{x}_f$ at the time t_f. Let $\check{\mathbf{x}}$ be the state on this trajectory at the time s where $t < s < t_f$.
Is the part $\check{\mathbf{x}}\mathbf{x}_\mathbf{f}$ of the trajectory $[\mathbf{x}, \overset{*}{\mathbf{u}}, \overset{*}{\mathbf{v}}]$ an optimal trajectory for a game that starts at $\check{\mathbf{x}}$?
The following lemma addresses this question.

Lemma 5.4.1 (Principle of optimality (weak))

Let the state $\mathbf{x}$ belong to the capture-set $\mathcal{W}$. Consider the optimal trajectory that starts at the state $\mathbf{x}$ at the time t and terminates on the target-set at the time t_f.
If $\check{\mathbf{x}}$ is the state on this trajectory at the time s where $t < s < t_f$ then:

$$J(\mathbf{x}) = \int_t^s L[\mathbf{x}(\tau), \overset{*}{\mathbf{u}}(\tau), \overset{*}{\mathbf{v}}(\tau)]\, d\tau + J(\check{\mathbf{x}}). \tag{5.3}$$

Proof
Let us denote:

$$k = \int_t^s L[\mathbf{x}(\tau), \overset{*}{\mathbf{u}}(\tau), \overset{*}{\mathbf{v}}(\tau)]\, d\tau.$$

(We shall denote a realization of an optimal strategy with the superscript $*$ and a realization that might involve a *deviation* from an optimal strategy with the superscript $\tilde{}$).
Now let us examine the outcomes of the following six plays:

Play 1: The play $\mathbf{x}\check{\mathbf{x}}\mathbf{x}_f$.
It starts at $\mathbf{x}$ at the time t and terminates at $\mathbf{x}_f$ at the time t_f. Both players realize their optimal strategies so $[\mathbf{x}, \overset{*}{\mathbf{u}}(\cdot), \overset{*}{\mathbf{v}}(\cdot)]$ is an optimal trajectory and the outcome, denoted as π_1, satisfies:

$$\pi_1 = \pi[\mathbf{x}, \overset{*}{\mathbf{u}}(\cdot), \overset{*}{\mathbf{v}}(\cdot)] = J(\mathbf{x}).$$

Play 2: The play $\check{\mathbf{x}}\mathbf{x}_F$.

It starts at $\check{\mathbf{x}}$ at the time s and terminates at $\mathbf{x}_F$ at the time t_F[4]. Both players realize their optimal strategies so $[\check{\mathbf{x}}, \overset{*}{\mathbf{u}}(\cdot), \overset{*}{\mathbf{v}}(\cdot)]$ is an optimal trajectory and the outcome, denoted as π_2, satisfies:

$$\pi_2 = \pi[\check{\mathbf{x}}, \overset{*}{\mathbf{u}}(\cdot), \overset{*}{\mathbf{v}}(\cdot)] = J(\check{\mathbf{x}}).$$

Play 3: The play $\check{\mathbf{x}}\mathbf{x}_{fp}$.

It starts at $\check{\mathbf{x}}$ at the time s. Player P realizes his optimal strategy $\overset{*}{\mathbf{u}}$, (which, by definition, guarantees termination), but player E realizes the strategy $\overset{*}{\mathbf{v}}$ that he used in play 1 (he assumes that the play started in $\mathbf{x}$ at the time t). The play terminates at $\mathbf{x}_{fp}$ at the time t_{fp}. The strategy used by player E might be a *deviation* from the optimal strategy for this game. So the outcome, denoted as π_3, satisfies:

$$\pi_3 = \pi[\check{\mathbf{x}}, \overset{*}{\mathbf{u}}(\cdot), \tilde{\mathbf{v}}(\cdot)] \leq J(\check{\mathbf{x}}).$$

Play 4: The play $\mathbf{x}\check{\mathbf{x}}\mathbf{x}_{fp}$.

The segment $\mathbf{x}\check{\mathbf{x}}$ of play 1 is compounded to the play $\check{\mathbf{x}}\mathbf{x}_{fp}$. This play starts at time t at $\mathbf{x}$ and terminates, as play 3, at time t_{fp} at $\mathbf{x}_{fp}$. Now player E's assumption that the game started at $\mathbf{x}$ is correct and his strategy is optimal. Player P on the other hand might have deviated from his optimal strategy! The outcome of this play, denoted as π_4, satisfies:

$$\pi_4 = \pi[\mathbf{x}, \tilde{\mathbf{u}}(\cdot), \overset{*}{\mathbf{v}}(\cdot)] \geq J(\mathbf{x}).$$

The relation between the outcomes of play 3 and play 4 is:

$$\pi_4 = \pi_3 + k.$$

Play 5: The play $\check{\mathbf{x}}\mathbf{x}_{fe}$

This is a play that starts at $\check{\mathbf{x}}$ at the time s. Player E plays optimally but player P uses the strategy he used in play 1[5] because he assumes that the play started at $\mathbf{x}$ and thus might have deviated from the optimal. The play terminates at the time t_{fe} at $\mathbf{x}_{fe}$. The outcome of this play, denoted as π_5, satisfies:

$$\pi_5 = \pi[\check{\mathbf{x}}, \tilde{\mathbf{u}}(\cdot), \overset{*}{\mathbf{v}}(\cdot)] \geq J(\check{\mathbf{x}}).$$

Play 6: The play $\mathbf{x}\check{\mathbf{x}}\mathbf{x}_{fe}$.

The segment $\mathbf{x}\check{\mathbf{x}}$ from play 1 is compounded with play 5. Here it is

[4] We do not assume that $\mathbf{x}_f = \mathbf{x}_F$ and that $t_f = t_F$.

[5] We have to argue that this strategy guarantees termination. The argument is left to the reader.

player E who might have deviated from the optimal!
The outcome of this play, denoted as π_6, satisfies

$$\pi_6 = \pi[\mathbf{x}, \overset{*}{\mathbf{u}}(\cdot), \tilde{v}(\cdot)] \leq J(\mathbf{x}).$$

The relation between the outcomes of play 5 and play 6 is:

$$\pi_6 = \pi_5 + k.$$

We therefore have:

$$\pi_5 + k = \pi_6 \leq J(\mathbf{x}) \leq \pi_4 = \pi_3 + k,$$

$$\pi_3 \leq J(\check{\mathbf{x}}) \leq \pi_5.$$

implying that:

$$\pi_3 = \pi_5 = J(\check{\mathbf{x}}),$$

And the lemma follows.

•

Lemma 5.4.2

Let

$$[\mathcal{W}, \mathcal{F}, \{\overset{*}{\mathbf{u}}_i\}, \{\overset{*}{\mathbf{v}}_i\}, J(\mathbf{x})],$$

be a solution of a differential game in the extended sense.
If an optimal trajectory starts at time t at a state $\mathbf{x}$ that belongs to the capture-set $\mathcal{W}$ and reaches a state $\check{\mathbf{x}}$ at the time s then we have for any given $\epsilon > 0$ that a number $n(\epsilon)$ exists such that for $i > n(\epsilon)$

$$J[\mathbf{x}(t)] = \int_t^s L[\mathbf{x}(\tau), \overset{*}{\mathbf{u}}_i(\tau), \overset{*}{\mathbf{v}}_i(\tau)]\, d\tau + J[\mathbf{x}(s)] + |t-s| O(\epsilon). \tag{5.4}$$

Proof
The proof is left to the reader.

•

5.4.2 Continuity of the Value Function along Optimal Trajectories

From Lemma 5.4.1 we can deduce the following very important property of the value function $J(\mathbf{x})$ along optimal trajectories:

Lemma 5.4.3

The value function $J(\mathbf{x})$ is continuous along optimal trajectories.

Proof
Let $\mathbf{x}(t)$ and $\mathbf{x}(t+dt)$ be two points on an optimal trajectory.
Let us apply Lemma 5.4.1:

$$J[\mathbf{x}(t)] = \int_t^{t+dt} L[\mathbf{x}(\tau), \overset{*}{\mathbf{u}}(\tau), \overset{*}{\mathbf{v}}(\tau)]d\tau + J[\mathbf{x}(t+dt)]. \tag{5.5}$$

We recall that the functions $\mathbf{f}(\mathbf{x}, \mathbf{u}, \mathbf{v})$ and $L(\mathbf{x}, \mathbf{u}, \mathbf{v})$ were both assumed to be bounded (see section 2.3 and section 2.6) and that the optimal strategy $\overset{*}{\mathbf{u}}$ guarantees termination for any trajectory that starts in the *capture-set*. This leads immediately to the following result:

$$\lim_{dt \to 0} J[\mathbf{x}(t+dt)] = J[\mathbf{x}(t)],$$

which completes the proof.

•

5.4.3 The Rate of Change of the Value Along Optimal Trajectories

We can now derive an important property regarding $\frac{d}{dt}J(\mathbf{x})$ along optimal trajectories[6].

Lemma 5.4.4

The rate of change of the value function $J(\mathbf{x})$ along an optimal trajectory satisfies:

$$\dot{J}(\mathbf{x}) = -L(\mathbf{x}, \overset{*}{\mathbf{u}}, \overset{*}{\mathbf{v}}). \tag{5.6}$$

proof.
The proof is a direct result of Lemma 5.4.3:

$$\lim_{dt \to 0} \frac{J[\mathbf{x}(t+dt)] - J[\mathbf{x}(t)]}{dt} = -L(\mathbf{x}, \overset{*}{\mathbf{u}}(\mathbf{x}), \overset{*}{\mathbf{v}}(\mathbf{x})). \tag{5.7}$$

•

relation (5.7) assures us that $\dot{J}(\mathbf{x})$ exists everywhere along an optimal trajectory (even at segments that are adjacent to a surface across which the value function is discontinuous!).

---

[6] $\dot{J}(\mathbf{x})$ is the rate with which the *isovalue surfaces* are traversed along an optimal trajectorey.

5.4.4 The Hamiltonian on Optimal Trajectories

We call points $\mathbf{x}$ where the gradient[7] $J_x(\mathbf{x})$ of the value function is continuous *regular* points. A segments of an optimal trajectories that consists only of *regular points* will be called a *regular part* of the optimal trtajectory.

The Hamiltonian function

At *regular* points we define the *Hamiltonian function* $H(\mathbf{x}, \mathbf{J}_x, \mathbf{u}, \mathbf{v})$ as:

$$H(\mathbf{x}, \mathbf{J}_x, \mathbf{u}, \mathbf{v}) = J_x \cdot \mathbf{f}(\mathbf{x}, \mathbf{u}, \mathbf{v}) + L(\mathbf{x}, \mathbf{u}, \mathbf{v}).$$

Lemma 5.4.5

On the regular parts of optimal trajectories we have:

$$H(\mathbf{x}, \mathbf{J}_x, \overset{*}{\mathbf{u}}, \overset{*}{\mathbf{v}}) = 0$$

Proof

The proof follows immediately from Lemma 5.4.4.

•

5.5 The Isaacs Equations

5.5.1 Slight Deviations From Optimal Strategies

The saddle inequalities (see relations (3.1)):

$$\pi[\mathbf{x}, \overset{*}{\mathbf{u}}(\cdot), \mathbf{v}(\cdot)] \leq \pi[\mathbf{x}, \overset{*}{\mathbf{u}}(\cdot), \overset{*}{\mathbf{v}}(\cdot)] \overset{E}{\preceq} \pi[\mathbf{x}, \mathbf{u}(\cdot), \overset{*}{\mathbf{v}}(\cdot)], \tag{5.8}$$

are comparison relations regarding two types of deviations from the optimal strategies, $[\mathbf{x}, \overset{*}{\mathbf{u}}(\cdot), \mathbf{v}(\cdot)]$ representing a play where player E deviated from his optimal strategy and $[\mathbf{x}, \mathbf{u}(\cdot), \overset{*}{\mathbf{v}}(\cdot)]$ representing a play where player P deviated from his optimal strategy.
The *deviations* represented in relation (5.8) are *global* in the sense that they relate to the whole play.

In the following we shall specialize relation (5.8) for cases where the play starts inside the capture-set ($\mathbf{x} \in \mathcal{W}$) and for deviations that are *slight* in the sense that the duration of the nonoptimal segment of the play is very short.

We shall define two types of *plays with slight deviations*:

[7] We shall use J_x as another notation for ∇J.

Type a

A play $[\mathbf{x}, \overset{*}{\mathbf{u}}(\cdot), \mathbf{v}(\cdot)]$ with a slight deviation by player E:
A play that starts at $\mathbf{x} \in \mathcal{W}$ and

$$\mathbf{v}(\cdot) \neq \overset{*}{\mathbf{v}}(\cdot) \quad \text{only for} \quad 0 < t \leq \epsilon_a,$$

$$\mathbf{v}(\cdot) = \overset{*}{\mathbf{v}}(\cdot) \quad \text{for} \quad \epsilon_a < t,$$

where[8]

$$\mathbf{x}(\epsilon_a) \in \mathcal{W},$$

and

$$\mathbf{x}(\epsilon_a) = \mathbf{x}(0) + \Delta_a \mathbf{x}.$$

The outcome of a play of type-a is:

$$\pi_a[\mathbf{x}, \overset{*}{\mathbf{u}}(\cdot), \mathbf{v}(\cdot)] = \int_0^{\epsilon_a} L[\mathbf{x}, \overset{*}{\mathbf{u}}(\tau), \mathbf{v}(\tau)]d\tau + J(\mathbf{x} + \Delta_a \mathbf{x}).$$

Type b

A play $[\mathbf{x}, \mathbf{u}(\cdot), \overset{*}{\mathbf{v}}(\cdot)]$ with a slight deviation by player P:
A play that starts at $\mathbf{x} \in \mathcal{W}$ and

$$\mathbf{u}(\cdot) \neq \overset{*}{\mathbf{u}}(\cdot) \quad \text{only for} \quad 0 < t \leq \epsilon_b,$$

$$\mathbf{u}(\cdot) = \overset{*}{\mathbf{u}}(\cdot) \quad \text{for} \quad \epsilon_b < t,$$

where ϵ_b is small so that in spite of the deviation by player P:

$$\mathbf{x}(\epsilon_b) \in \mathcal{W},$$

and

$$\mathbf{x}(\epsilon_b) = \mathbf{x}(0) + \Delta_b \mathbf{x}.$$

The outcome of a play of type-b is:

$$\pi_b[\mathbf{x}, \mathbf{u}(\cdot), \overset{*}{\mathbf{v}}(\cdot)] = \int_0^{\epsilon_b} L[\mathbf{x}, \mathbf{u}(\tau), \overset{*}{\mathbf{v}}(\tau)]d\tau + J(\mathbf{x} + \Delta_b \mathbf{x}).$$

The saddle property (5.8) provides that:

$$\pi_a[\mathbf{x}, \overset{*}{\mathbf{u}}(\cdot), \mathbf{v}(\cdot)] \leq J(\mathbf{x}) \leq \pi_b[\mathbf{x}, \mathbf{u}(\cdot), \overset{*}{\mathbf{v}}(\cdot)],$$

---

[8] Recall that $\overset{*}{U}$ is a strategy that guarantees termination.

or:

$$\int_0^{\epsilon_a} L[\mathbf{x}, \overset{*}{\mathbf{u}}, \mathbf{v}]d\tau + J(\mathbf{x} + \Delta_a \mathbf{x}) \le J(\mathbf{x}) \le$$

$$\int_0^{\epsilon_b} L[\mathbf{x}, \mathbf{u}, \overset{*}{\mathbf{v}}]d\tau + J(\mathbf{x} + \Delta_b \mathbf{x}). \tag{5.9}$$

The general saddle inequality (relations (3.1)) is a *global* criterion. Relation (5.9), on the other hand, is of *semilocal* nature.
Next we shall use relation (5.9) to derive a *local* qualification of optimality.

5.5.2 The Isaacs Main Equations

Isaacs' equations are probably the most important property of the solutions of differential games. The Isaacs' equations can be considered as an extension to the well known *Hamilton-Jacobi-Bellman-Pontriyagin minimum principle* in the theory of optimal control.
We shall distinguish between the *regular part* of an optimal trajectory, where $\nabla J(\mathbf{x})$ is continuous, and the *singular* part of the optimal trajectory where discontinuities in either $J(\mathbf{x})$ or $\nabla J(\mathbf{x})$ occur.
In this subsection we shall discuss the Isaacs' theorems for the regular parts of an optimal trajectory.

Theorem 5.5.1 (The Main Equations of Isaacs)

For points $\mathbf{x}$ in the capture-set $\mathcal{W}$ that belong to *regular* parts of optimal trajectories the following relations must hold:

Main Equation 1 (ME1)

$$\underset{\forall \mathbf{v} \in \mathcal{V}}{H(\mathbf{x}, \mathbf{J}_x, \overset{*}{\mathbf{u}}, \mathbf{v})} \le H(\mathbf{x}, \mathbf{J}_x, \overset{*}{\mathbf{u}}, \overset{*}{\mathbf{v}}) \le \underset{\forall \mathbf{u} \in \mathcal{U}}{H(\mathbf{x}, \mathbf{J}_x, \mathbf{u}, \overset{*}{\mathbf{v}})}. \tag{5.10}$$

Main Equation 2 (ME2)

$$H(\mathbf{x}, \mathbf{J}_x, \overset{*}{\mathbf{u}}, \overset{*}{\mathbf{v}}) = 0. \tag{5.11}$$

Proof
We do not have to prove ME2, (relation (5.11)), because it is identical to Lemma 5.4.5.

Let us proceed and prove ME1:
For sufficiently small Δt we have from relation (5.9) that the following will be true:

$$J(\mathbf{x} + \Delta\mathbf{x}) + \int_0^{\Delta t} L[\mathbf{x}(\tau), \overset{*}{\mathbf{u}}(\tau), \mathbf{v}(\tau)]d\tau \le J(\mathbf{x}). \tag{5.12}$$

Let us recall that:

1. $\mathbf{x}$ belongs to a *regular* part of an optimal trajectory.
2. $L(\mathbf{x}, \mathbf{u}, \mathbf{v})$ was assumed to be bounded. (See section 2.3).
3. $\overset{*}{\mathbf{U}}$ is a strategy that guarantees termination in finite time.
4. As $\Delta t \to 0$ also $|\Delta \mathbf{x}| \to 0$.
5. $\lim_{\Delta t \to 0} \frac{\Delta \mathbf{x}}{\Delta t} = \mathbf{f}(\mathbf{x}, \overset{*}{\mathbf{u}}, \mathbf{v})$

Evidently relation (5.12) can now be rewritten in a differential form:

$$\begin{aligned} &\nabla J(\mathbf{x}) \cdot \mathbf{f}(\mathbf{x}, \overset{*}{\mathbf{u}}, \mathbf{v}) + \mathbf{L}(\mathbf{x}, \overset{*}{\mathbf{u}}, \mathbf{v}) \leq \mathbf{0}, \\ &\forall \mathbf{v} \in \mathcal{V} \end{aligned} \tag{5.13}$$

which is the same as

$$H(\mathbf{x}, \mathbf{J}_x, \overset{*}{\mathbf{u}}, \mathbf{v}) \leq 0 \quad \text{for all} \quad \mathbf{v} \in \mathcal{V}. \tag{5.14}$$

This completes the proof of the left hand side of relation (5.10).

The proof that

$$0 \leq H(\mathbf{x}, \mathbf{J}_x, \mathbf{u}, \overset{*}{\mathbf{v}}), \tag{5.15}$$

is similar and is left to the reader.

•

Corollary 5.5.2

In terminal cost games the game-surfaces[9] $J(\mathbf{x}) = k$ are semipermeable surfaces.

Proof

The proof follows immediately from relation (5.16) and from the definition of semipermability of surfaces (see relation 4.2).

•

Let us try to interpret Isaacs' equations.
A very illustrative interpretation was suggested by *G. Leitmann* and *A. Blaquiere* in [ABL69]. They treated a generalized differential game with an outcome that consists only of a terminal payoff ($L = 0$). (We saw in

---

[9]The name *game surfaces* was given to the surfaces $J(\mathbf{x}) = k$ by *G. Leitmann* and *A. Blaquiere* in [ABL69].

Lemma 2.6.1 that any game can be reformulated as a *terminal cost game*). For this case the Isaacs' theorems have the form:

$$\underset{\forall \mathbf{v} \in \mathcal{V}}{\nabla J(\mathbf{x}) \cdot \mathbf{f}(\mathbf{x}, \overset{*}{\mathbf{u}}, \mathbf{v})} \leq 0 = \nabla J(\mathbf{x}) \cdot \mathbf{f}(\mathbf{x}, \overset{*}{\mathbf{u}}, \overset{*}{\mathbf{v}}) \leq \underset{\forall \mathbf{u} \in \mathcal{U}}{\nabla J(\mathbf{x}) \cdot \mathbf{f}(\mathbf{x}, \mathbf{u}, \overset{*}{\mathbf{v}})}. \tag{5.16}$$

As long as both players play optimally the trajectory remains on the same game-surface. If player P deviates while player E continues to play $\overset{*}{\mathbf{v}}$ the trajectory ascends to a higher *game surface* and if player E deviates while player P continue to play $\overset{*}{\mathbf{u}}$ the trajectory descends to a lower *game surface*.

We can thus view the Hamiltonians $H(\mathbf{x}, \mathbf{J}_x, \mathbf{u}, \overset{*}{\mathbf{v}})$ and $H(\mathbf{x}, \mathbf{J}_x, \overset{*}{\mathbf{u}}, \mathbf{v})$ as measures of the deviations (ascend or descend) from the ideal $J(\mathbf{x})$.

•

Remarks

Two remarks are in due place here;

Nonadmissible strategies

The reader should notice that Isaacs' theorems are valid also for a solution where the optimal strategies do not satisfy the *infornmation constraints*.

Seperable Hamiltonian functions

In many problems we have a model with a *separable* Hamiltoian function:

$$H(\mathbf{x}, J_x, \mathbf{u}, \mathbf{v}) = H_p(\mathbf{x}, J_x, \mathbf{u}) + H_e(\mathbf{x}, J_x, \mathbf{v}).$$

Whenever we write Isaac's theorem (ME1) as:

$$H(\mathbf{x}, \mathbf{J}_x, \overset{*}{\mathbf{u}}, \overset{*}{\mathbf{v}}) = \min_{\mathbf{u} \in \mathcal{U}} \max_{\mathbf{v} \in \mathcal{V}} H(\mathbf{x}, \mathbf{J}_x, \mathbf{u}, \mathbf{v}), \tag{5.17}$$

we mean that the Hamiltonian functions is *separable* and that

$$H(\mathbf{x}, \mathbf{J}_x, \overset{*}{\mathbf{u}}, \overset{*}{\mathbf{v}}) = \min_{\mathbf{u} \in \mathcal{U}} H_p(\mathbf{x}, J_x, \mathbf{u}) + \max_{\mathbf{v} \in \mathcal{V}} H_e(\mathbf{x}, J_x, \mathbf{v}). \tag{5.18}$$

5.5.3 The hodograph representations of ME1

We can express Isaacs' equation ME1 in an alternative form where we replace relation (5.10) by two *optimization problems*:

1. $\overset{*}{\mathbf{v}}$ solves the problem:

$$\max_{\mathbf{v} \in \mathcal{V}} H(\mathbf{x}, \mathbf{J}_x, \overset{*}{\mathbf{u}}, \mathbf{v}) \tag{5.19}$$

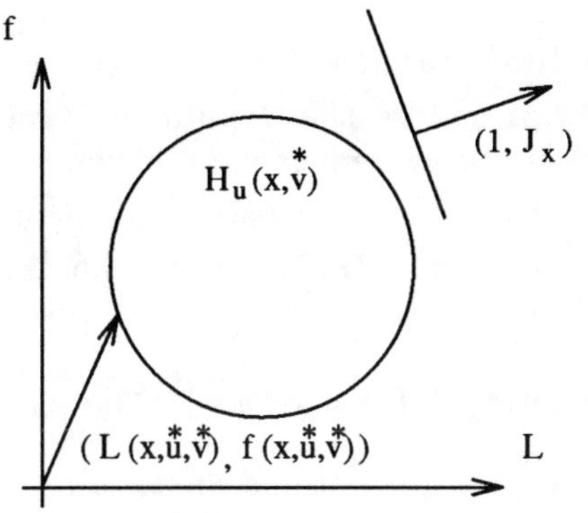

FIGURE 5.1. A hodograph representation of ME1

2. $\overset{*}{\mathbf{u}}$ solves the problem:

$$\max_{\mathbf{u}\in\mathcal{U}} H(\mathbf{x}, \mathbf{J}_x, \mathbf{u}, \overset{*}{\mathbf{v}}) \tag{5.20}$$

It is illustrative to have a geometric characterization of the optimization problems (5.20) and (5.19). This is done by using the *hodograph representation* (see Figure 5.1).

The hodograph-set $\mathcal{H}_u(\mathbf{x}, \overset{*}{\mathbf{v}})$ is a set of (n+1)-dimensional vectors defined as:

$$\mathcal{H}_u(\mathbf{x}, \overset{*}{\mathbf{v}}) \equiv \{\ (\mathbf{f}(\mathbf{x}, \mathbf{u}, \overset{*}{\mathbf{v}}), \mathbf{L}(\mathbf{x}, \mathbf{u}, \overset{*}{\mathbf{v}}))\ |\ \mathbf{u} \in \mathcal{U}\}.$$

Similarly, the hodograph-set $\mathcal{H}_v(\mathbf{x}, \overset{*}{\mathbf{v}})$ is a set of (n+1)-dimensional vectors defined as:

$$\mathcal{H}_v(\mathbf{x}, \overset{*}{\mathbf{u}}) \equiv \{\ [\mathbf{f}(\mathbf{x}, \overset{*}{\mathbf{u}}, \mathbf{v}), \mathbf{L}(\mathbf{x}, \overset{*}{\mathbf{u}}, \mathbf{v})]\ |\ \mathbf{v} \in \mathcal{V}\ \}.$$

The hodograph-set is considered to be *regular* if it is spanned by (n+1) base vectors and if $\nabla J(\mathbf{x}) \neq 0$. In the following discussion, unless specified otherwise, we assume that the hodograph representations are *regular*.

Lemma 5.5.3

The hodograph-sets $\mathcal{H}_v(\mathbf{x}, \overset{*}{\mathbf{u}})$ and $\mathcal{H}_u(\mathbf{x}, \overset{*}{\mathbf{v}})$ are *convex*.

Proof

The proof is based on the assumed convexity of the functions $\mathbf{f}(\mathbf{x}, \overset{*}{\mathbf{u}}, \mathbf{v})$ and $L(\mathbf{x}, \overset{*}{\mathbf{u}}, \mathbf{v})$ (see section 2.3 and section 2.6). The details are left to the reader.

•

Remark.
We did not demand that the functions $\mathbf{f}(\mathbf{x}, \overset{*}{\mathbf{u}}, \mathbf{v})$ and $L(\mathbf{x}, \overset{*}{\mathbf{u}}, \mathbf{v})$ should be *strictly convex* in $\mathbf{u}$ and $\mathbf{v}$. The Hamiltonian function $H(\mathbf{x}, \mathbf{J}_x, \overset{*}{\mathbf{u}}, \mathbf{v})$, for example, can be linear in component(s) of the control $\mathbf{v}$ (or similarly, $H(\mathbf{x}, \mathbf{J}_x, \mathbf{u}, \overset{*}{\mathbf{v}})$ can be linear in component(s) of the control $\mathbf{u}$).
If the Hamiltonian function $H(\mathbf{x}, \mathbf{J}_x, \mathbf{u}, \overset{*}{\mathbf{v}})$ is linear in one component of the control $\mathbf{u}$ say in u_1.

$$H(\mathbf{x}, \mathbf{J}_x, \mathbf{u}, \overset{*}{\mathbf{v}}) = B[\mathbf{x}, J_x, \mathbf{u}_2, \overset{*}{\mathbf{v}}] + A[\mathbf{x}, J_x, \mathbf{u}_2, \overset{*}{\mathbf{v}}]u_1,$$

then by ME1 we have that the optimal choice of u_1 is:

$$\begin{aligned} \overset{*}{u}_1 &= \operatorname{sign} A(\mathbf{x}, J_x, \mathbf{u}_2, \overset{*}{\mathbf{v}}) \qquad \text{for } A(\mathbf{x}, J_x, \mathbf{u}_2, \overset{*}{\mathbf{v}}) \neq 0, \\ \overset{*}{u}_1 &= \text{any admissible } u_1 \qquad \text{for } A(\mathbf{x}, J_x, \mathbf{u}_2, \overset{*}{\mathbf{v}}) = 0. \end{aligned} \tag{5.21}$$

The function $A(\mathbf{x}, J_x, \mathbf{u}_2, \overset{*}{\mathbf{v}})$ is called the *switch function* for the control u_1. The corresponding hodograph-set $\mathcal{H}_u(\mathbf{x}, \overset{*}{\mathbf{v}})$ is *convex* but not *strictly convex*.

•

We may now state ME1 in the terms of the hodograph representation as:

Corollary 5.5.4 (The hodograph equivalents of ME1)

The vector $[\mathbf{f}(\mathbf{x}, \overset{*}{\mathbf{u}}, \overset{*}{\mathbf{v}}), L(\mathbf{x}, \overset{*}{\mathbf{u}}, \overset{*}{\mathbf{v}})]$ is on the boundary of the hodograph-sets $\mathcal{H}_u(\mathbf{x}, \overset{*}{\mathbf{v}})$ and $\mathcal{H}_v(\mathbf{x}, \overset{*}{\mathbf{u}})$ and solves the optimization problem:

$$\begin{aligned} &\min\{(\mathbf{f}(\mathbf{x}, \mathbf{u}, \overset{*}{\mathbf{v}}), \mathbf{L}(\mathbf{x}, \mathbf{u}, \overset{*}{\mathbf{v}})) \cdot [\mathbf{J_X}(\mathbf{x}), \mathbf{1}]\} \\ &(\mathbf{f}(\mathbf{x}, \mathbf{u}, \overset{*}{\mathbf{v}}), L(\mathbf{x}, \mathbf{u}, \overset{*}{\mathbf{v}})) \in \mathcal{H}_u(\mathbf{x}, \overset{*}{\mathbf{v}}) \end{aligned}$$

and the optimization problem:

$$\begin{aligned} &\max\{(\mathbf{f}(\mathbf{x}, \overset{*}{\mathbf{u}}, \mathbf{v}), \mathbf{L}(\mathbf{x}, \overset{*}{\mathbf{u}}, \mathbf{v})) \cdot [\mathbf{J_X}(\mathbf{x}), \mathbf{1}]\} \\ &(\mathbf{f}(\mathbf{x}, \overset{*}{\mathbf{u}}, \mathbf{v}), \mathbf{L}(\mathbf{x}, \overset{*}{\mathbf{u}}, \mathbf{v})) \in \mathcal{H}_{\mathbf{V}}(\mathbf{x}, \overset{*}{\mathbf{u}}) \end{aligned}$$

•

From the property of the convexity of the hodograph-sets, (see Figure 5.2), we can infer that

Corollary 5.5.5

The solution of the optimization problem

$$\min_{\mathbf{u} \in \mathcal{U}} H(\mathbf{x}, \mathbf{J}_x, \mathbf{u}, \overset{*}{\mathbf{v}})$$

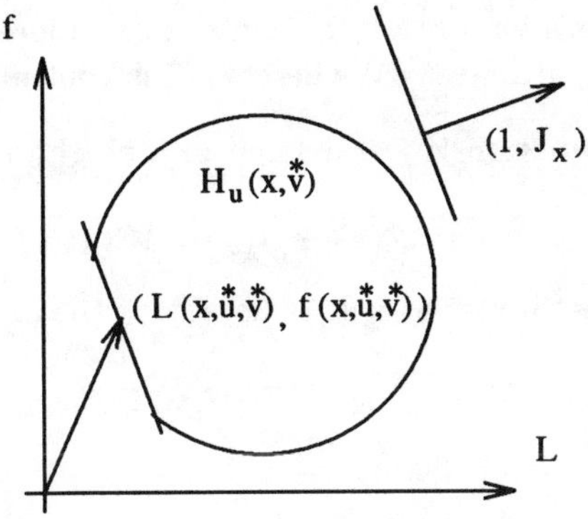

FIGURE 5.2. A case with nonunique solution

may be *nonunique* only if the Hamiltonian function $H(\mathbf{x}, \mathbf{J}_x, \mathbf{u}, \overset{*}{\mathbf{v}})$ is linear in components of the control $\mathbf{u}$. The solution of the optimization problem

$$\max_{\mathbf{v}\in\mathcal{V}} H(\mathbf{x}, \mathbf{J}_x, \overset{*}{\mathbf{u}}, \mathbf{v})$$

may be *nonunique* only if the Hamiltonian function $H(\mathbf{x}, \mathbf{J}_x, \overset{*}{\mathbf{u}}, \mathbf{v})$ is linear in components of the control $\mathbf{v}$.

•

5.5.4 The Viscosity Form of Isaacs Equations

In section 5.5.2 we presented the classical *Isaacs equations* which constitute a local necessary condition for optimality at points $\mathbf{x}$ where $J(\mathbf{x})$ is continuous and $\nabla J(\mathbf{x})$ exists. In this section we give the *viscosity form of Isaacs equations*[10] which is a local necessary condition to be met at *all* the points $\mathbf{x}$ where $J(\mathbf{x})$ is continuous.

Theorem 5.5.6 (Viscosity Conditions)

---

[10]It was *P. L. Lions* who first noticed, in the context of control theory, that the Hamilton-Jacobi-Bellmann-Isaacs equation holds in the *viscosity sense*. *P. E. Souganidis* , *M. Bardi* and *P. Soravia* extended the use of the notion of *viscosity solutions* of the Isaacs' equation in investigating the questions of existence and uniqueness for various classes of differential games. (See Lions and Souganidis in [Ls85], Bardi and Soravia in [BS89] and Bernhard in [Ber87]).

Let $[\mathcal{W}, \mathcal{F}, \overset{*}{\mathbf{u}}, \overset{*}{\mathbf{v}}, J(\mathbf{x})]$ be a solution quintet of a differential game and let $\mathbf{x}$ be a point in the capture-set $\mathcal{W}$ where $J(\mathbf{x})$ is continuous. Then

1. For any C^1 function $W(\mathbf{x})$ defined in a neighborhood of $\mathbf{x}$ such that the function
$$Z(\mathbf{x}) = J(\mathbf{x}) - W(\mathbf{x}),$$
has a local maximum at $\mathbf{x}$ It is necessary that:
$$\begin{array}{c} H(\mathbf{x}, W_x, \mathbf{u}, \overset{*}{\mathbf{v}}) \geq 0. \\ \forall \mathbf{u} \in \mathcal{U} \end{array} \tag{5.22}$$

2. For any C^1 function $W(\mathbf{x})$ defined in a neighborhood of $\mathbf{x}$ such that the function
$$Z(\mathbf{x}) = J(\mathbf{x}) - W(\mathbf{x}),$$
has a local minimum at $\mathbf{x}$ It is necessary that:
$$\begin{array}{c} H(\mathbf{x}, W_x, \overset{*}{\mathbf{u}}, \mathbf{v}) \leq 0. \\ \forall \mathbf{v} \in \mathcal{V} \end{array} \tag{5.23}$$

Proof
Let us first prove part 1 of the theorem.
With no loss of generality it suffices to prove the theorem for the cases where $W(\mathbf{x})$ be any C^1 function defined in a neighborhood $\mathcal{H}(\mathbf{x})$ of $\mathbf{x}$ such that
$$Z(\mathbf{x}) = J(\mathbf{x}) - W(\mathbf{x}) = 0, \tag{5.24}$$
and for all $\mathbf{x} + \Delta\mathbf{x} \in \mathcal{H}(\mathbf{x})$,
$$Z(\mathbf{x} + \Delta\mathbf{x}) = J(\mathbf{x} + \Delta\mathbf{x}) - W(\mathbf{x} + \Delta\mathbf{x}) \leq 0. \tag{5.25}$$

Let us consider the outcome π_b of a play that starts at $\mathbf{x}$ at the time t and belongs to the following class of plays with *slight deviation* of *type b*:[11]. Player E plays $\overset{*}{\mathbf{v}}$ throughout the whole play. Player P may deviate from his optimal strategy $\overset{*}{\mathbf{u}}$ during the time interval $t < \tau \leq t + \Delta t$ and proceed optimally thereafter where
$$\mathbf{x}(t + \Delta t) = \mathbf{x} + \Delta\mathbf{x} \in \mathcal{H}(\mathbf{x}).$$

From the *saddle property* (5.8) we have for the outcome π_b of this play:
$$J(\mathbf{x}) \leq \pi_b = \int_t^{t+\Delta t} L(\mathbf{x}, \mathbf{u}, \overset{*}{\mathbf{v}})\, d\tau + J(\mathbf{x} + \Delta\mathbf{x}). \tag{5.26}$$

[11] Recall section 5.5.1.

We have assumed that $\mathbf{f}(\mathbf{x}, \mathbf{u}, \mathbf{v})$ and $L(\mathbf{x}, \mathbf{u}, \mathbf{v})$ are bounded and that $J(\mathbf{x})$ is continuous at $\mathbf{x}$ so there exists a constant $\gamma > 0$ such that:

$$\mathbf{x} + \Delta\mathbf{x} \in \mathcal{H}(\mathbf{x}) \quad \text{for all} \quad \Delta t < \gamma. \tag{5.27}$$

Substituting

$$J(\mathbf{x}) = W(\mathbf{x} + \Delta\mathbf{x}) - [W(\mathbf{x} + \Delta\mathbf{x}) - W(\mathbf{x})],$$

we obtain for this class of plays that for all $\Delta t < \gamma$,

$$\int_t^{t+\Delta t} L(\mathbf{x}, \mathbf{u}, \overset{*}{\mathbf{v}})d\tau + W(\mathbf{x}+\Delta\mathbf{x}) - W(\mathbf{x}) \geq W(\mathbf{x}+\Delta\mathbf{x}) - J(\mathbf{x}+\Delta\mathbf{x}). \tag{5.28}$$

$W(\mathbf{x})$ is assumed to be a C^1 function in $\mathcal{H}(\mathbf{x})$ so by letting $\Delta t \to 0$ and recalling relation (5.25) we see that the differential equivalent of relation (5.28) satisfies:

$$\begin{array}{l} H(\mathbf{x}, W_x, \mathbf{u}, \overset{*}{\mathbf{v}}) \geq 0. \\ \forall \mathbf{u} \in \mathcal{U} \end{array} \tag{5.29}$$

This proves the claim of part 1 of the theorem.

The proof for part 2 of the theorem is similar and is left to the reader.

•

5.6 The Adjoint Equations

When we discuss the *equations of motion* $\dot{\mathbf{x}} = \mathbf{f}(\mathbf{x}, \mathbf{u}, \mathbf{v})$ we often talk of two groups of independent variables:

1. *State variables*
   The n components of the *state vector* $\mathbf{x}$,

2. *Control variables*
   The control vectors $\mathbf{u}$ and $\mathbf{v}$.

Observing the *Hamiltonian* function $H(\mathbf{x}, \mathbf{J}_x, \mathbf{u}, \mathbf{v})$ we find that it also depends on an additional group of variables, the n components of J_x, called *adjoint variables*.

In this section we shall describe an important necessary condition regarding the time rate of change of the *adjoint variables* along *regular* parts of optimal trajectories.

Let us define abbreviations for the following functions:

$$
\begin{aligned}
H_x &\equiv \frac{\partial}{\partial x} H(\mathbf{x}, \mathbf{J}_x, \overset{*}{\mathbf{u}}, \overset{*}{\mathbf{v}}),\\
J_{xx} &\equiv \frac{\partial}{\partial x} J_x(\mathbf{x}),\\
H_{J_x} &\equiv \frac{\partial}{\partial J_x} H(\mathbf{x}, \mathbf{J}_x, \overset{*}{\mathbf{u}}, \overset{*}{\mathbf{v}}),\\
H_u &\equiv \frac{\partial}{\partial \overset{*}{\mathbf{u}}} H(\mathbf{x}, \mathbf{J}_x, \overset{*}{\mathbf{u}}, \overset{*}{\mathbf{v}}),\\
H_v &\equiv \frac{\partial}{\partial \overset{*}{\mathbf{v}}} H(\mathbf{x}, \mathbf{J}_x, \overset{*}{\mathbf{u}}, \overset{*}{\mathbf{v}}),\\
\overset{*}{\mathbf{u}}_x &\equiv \frac{\partial}{\partial x} \overset{*}{\mathbf{u}}(\mathbf{x}),\\
\overset{*}{\mathbf{v}}_x &\equiv \frac{\partial}{\partial x} \overset{*}{\mathbf{v}}(\mathbf{x}),\\
\overset{*}{\mathbf{f}} &\equiv \mathbf{f}(\mathbf{x}, \overset{*}{\mathbf{u}}, \overset{*}{\mathbf{v}}).
\end{aligned}
\tag{5.30}
$$

Theorem 5.6.1 (The adjoint equations)

It is necessary that *along* regular parts of optimal trajectories:

$$
\frac{d}{dt} J_x[\mathbf{x}(t)] = -\frac{\partial}{\partial x} H(\mathbf{x}, \mathbf{J}_x, \overset{*}{\mathbf{u}}, \overset{*}{\mathbf{v}}). \tag{5.31}
$$

Proof

It follows from the definition that if $\mathbf{x}$ is a *regular point* of an optimal trajectory then all the points in some neighborhood of $\mathbf{x}$ are also *regular points* and that for sufficiently small $\Delta\mathbf{x}$:

$$
\begin{aligned}
&H[\mathbf{x}, J_x(\mathbf{x}), \overset{*}{\mathbf{u}}(\mathbf{x}), \overset{*}{\mathbf{v}}(\mathbf{x})] =\\
&H[(\mathbf{x}+\Delta\mathbf{x}), J_x(\mathbf{x}+\Delta\mathbf{x}), \overset{*}{\mathbf{u}}(\mathbf{x}+\Delta\mathbf{x}), \overset{*}{\mathbf{v}}(\mathbf{x}+\Delta\mathbf{x})] = 0.
\end{aligned}
\tag{5.32}
$$

Since relation (5.32) is valid for any arbitrary small $\Delta\mathbf{x}$ we can express it in a *differential form*:

$$
H_x + J_{xx} \cdot \overset{*}{\mathbf{f}} + H_u \cdot \overset{*}{\mathbf{u}}_x + H_v \cdot \overset{*}{\mathbf{v}}_x = 0. \tag{5.33}
$$

In order to prove the theorem it suffices to prove that:

$$
\begin{aligned}
H_u \cdot \overset{*}{\mathbf{u}}_x &= 0,\\
H_v \cdot \overset{*}{\mathbf{v}}_x &= 0.
\end{aligned}
\tag{5.34}
$$

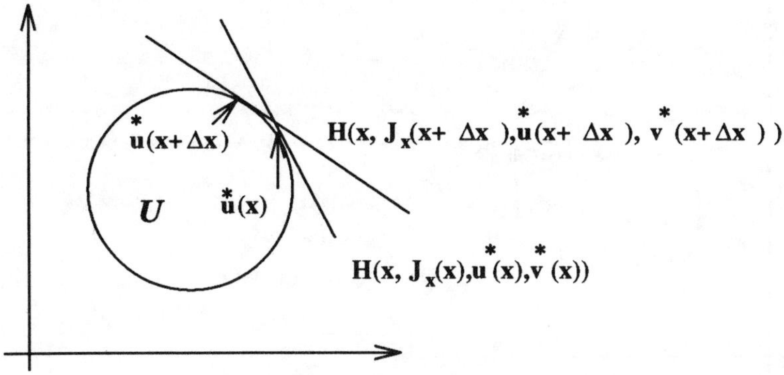

FIGURE 5.3. $\overset{*}{\mathbf{u}}$ on a smooth boundary of $\mathcal{U}$

We shall give a heuristic description of the proof that $H_u \cdot \overset{*}{\mathbf{u}}_x = 0$[12] by investigating the functional properties of the optimization problem

$$\min_{\mathbf{u} \in \mathcal{U}} H(\mathbf{x}, \mathbf{J}_x, \mathbf{u}, \overset{*}{\mathbf{v}}) = H(\mathbf{x}, \mathbf{J}_x, \overset{*}{\mathbf{u}}, \overset{*}{\mathbf{v}}). \tag{5.35}$$

We shall sketch the proof of the assertion $H_u \cdot \overset{*}{\mathbf{u}}_x = 0$, for the three possible different cases concerning the location of the solution of the optimization problem, the vector $\overset{*}{\mathbf{u}}$, in the set $\mathcal{U}$ of admissible controls.

Case 1
$\overset{*}{\mathbf{u}}$ is in the *interior* of the set $\mathcal{U}$.
In this case the well known necessary condition for $\overset{*}{\mathbf{u}}$ to be a solution of (5.35) is:

$$H_u(\mathbf{x}, J_x, \overset{*}{\mathbf{u}}, \overset{*}{\mathbf{v}}) = 0, \tag{5.36}$$

and $H_u \cdot \overset{*}{\mathbf{u}}_x = 0$ is assured.

Case 2
$\overset{*}{\mathbf{u}}$ belongs to a smooth part of the boundary of the set $\mathcal{U}$, (see Figure 5.3).

We have that $\overset{*}{\mathbf{u}}\ (\mathbf{x})$ solves

$$\min_{\mathbf{u} \in \mathcal{U}} H(\mathbf{x}, \mathbf{J}_x, \mathbf{u}, \overset{*}{\mathbf{v}}) = H(\mathbf{x}, \mathbf{J}_x, \overset{*}{\mathbf{u}}, \overset{*}{\mathbf{v}}) = 0,$$

and that the surface $H(\mathbf{x}, \mathbf{J}_x, \overset{*}{\mathbf{u}}, \overset{*}{\mathbf{v}}) = 0$ in the $\mathbf{u}$-space is tangent to the boundary of the set $\mathcal{U}$ at $\overset{*}{\mathbf{u}}$.

---

[12]The proof that $H_v \cdot \overset{*}{\mathbf{v}}_x = 0$ will be left to the reader.

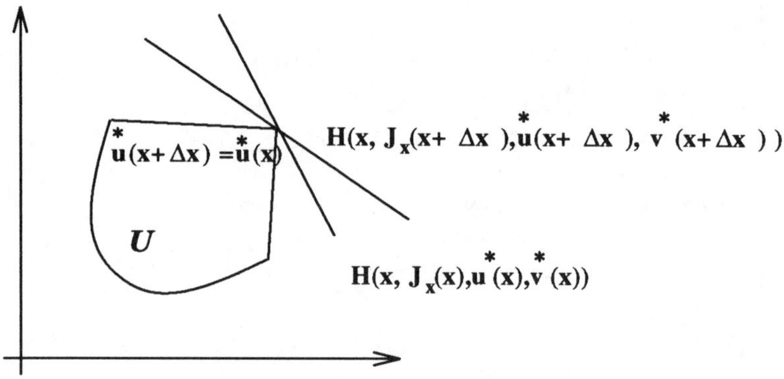

FIGURE 5.4. $\overset{*}{\mathbf{u}}$ on a corner on the boundary of $\mathcal{U}$

If $\overset{*}{\mathbf{u}}\,(\mathbf{x}+\mathbf{\Delta x})$, that solves

$$\min_{\mathbf{u}\in\mathcal{U}} H[\mathbf{x}+\mathbf{\Delta x}, J_x, \mathbf{u}, \overset{*}{\mathbf{v}}] = H[\mathbf{x}+\mathbf{\Delta x}, J_x, \overset{*}{\mathbf{u}}, \overset{*}{\mathbf{v}}] = 0,$$

is also on the boundary of $\mathcal{U}$ then the surface

$$H[\mathbf{x}+\mathbf{\Delta x}, J_x(\mathbf{x}+\mathbf{\Delta x}), \overset{*}{\mathbf{u}}\,(\mathbf{x}+\mathbf{\Delta x}), \overset{*}{\mathbf{v}}\,(\mathbf{x}+\mathbf{\Delta x})] = 0,$$

is also tangent to the set $\mathcal{U}$ at its boundary at the point $\overset{*}{\mathbf{u}}\,(\mathbf{x}+\mathbf{\Delta x})$.
In this case $\overset{*}{\mathbf{u}}\,(\mathbf{x}) \neq \overset{*}{\mathbf{u}}\,(\mathbf{x}+\mathbf{\Delta x})$, but obviously

$$lim_{|\mathbf{\Delta x}|\to 0}\frac{\overset{*}{\mathbf{u}}(\mathbf{x}+\mathbf{\Delta x})-\overset{*}{\mathbf{u}}(\mathbf{x})}{|\Delta x|} \text{ is tangent to } \mathcal{U} \text{ at } \overset{*}{\mathbf{u}}\,(\mathbf{x}).$$

It follows that $\overset{*}{\mathbf{u}}_x\,(\mathbf{x})$ lies on the surface $H(\mathbf{x}, \mathbf{J}_x, \overset{*}{\mathbf{u}}, \overset{*}{\mathbf{v}}) = 0$ which is *tangent* to the set $\mathcal{U}$ at $\overset{*}{\mathbf{u}}\,(\mathbf{x})$ and is hence *orthogonal* to $H_u[\mathbf{x}, J_x(\mathbf{x}), \overset{*}{\mathbf{u}}\,(\mathbf{x}), \overset{*}{\mathbf{u}}\,(\mathbf{x})]$ and $H_u \cdot \overset{*}{\mathbf{u}}_x = 0$ follows.

•

Case 3
$\overset{*}{\mathbf{u}}$ belongs to a *corner* on the boundary of the set $\mathcal{U}$, (see Figure 5.4).
We have that $\overset{*}{\mathbf{u}}\,(\mathbf{x})$ solves

$$\min_{\mathbf{u}\in\mathcal{U}} H(\mathbf{x}, \mathbf{J}_x, \mathbf{u}, \overset{*}{\mathbf{v}}) = H(\mathbf{x}, \mathbf{J}_x, \overset{*}{\mathbf{u}}, \overset{*}{\mathbf{v}}) = 0,$$

and that the surface $H(\mathbf{x}, \mathbf{J}_x, \overset{*}{\mathbf{u}}, \overset{*}{\mathbf{v}}) = 0$, in the $\mathbf{u}$-space, is tangent to the corner of the boundary of the set $\mathcal{U}$ at $\overset{*}{\mathbf{u}}$.
If $\overset{*}{\mathbf{u}}\,(\mathbf{x}+\mathbf{\Delta x})$, that solves

$$\min_{\mathbf{u}\in\mathcal{U}} H[\mathbf{x}+\mathbf{\Delta x}, J_x, \mathbf{u}, \overset{*}{\mathbf{v}}] = H[\mathbf{x}+\mathbf{\Delta x}, J_x, \overset{*}{\mathbf{u}}, \overset{*}{\mathbf{v}}] = 0,$$

is also on the corner of the boundary of $\mathcal{U}$, then

$$\overset{*}{\mathbf{u}}(\mathbf{x}) = \overset{*}{\mathbf{u}}(\mathbf{x} + \mathbf{\Delta x}),$$

and

$$lim_{|\mathbf{\Delta x}| \to 0} \frac{\overset{*}{\mathbf{u}}(\mathbf{x} + \mathbf{\Delta x}) - \overset{*}{\mathbf{u}}(\mathbf{x})}{|\Delta x|} = 0.$$

So $\overset{*}{\mathbf{u}}_x(\mathbf{x}) = 0$ and the assertion $H_u \cdot \overset{*}{\mathbf{u}}_x = 0$ is again true.

•

The adjoint equations (5.31) are a key tool for constructing candidate optimal trajectories. However we should never forget that the differential formulation (5.33) is valid *only* for regular parts of optimal trajectories. In a later chapter (see section 10.6), we shall use a modified use of relation (5.32) to derive *adjoint equations* along singular arcs.

5.6.1 The Retro Time Form of the Adjoint Equations

The variable t in equation (5.31) has the meaning of *forward-time* in the sense that we associate the time $t = 0$ to the starting point of the optimal trajectory and the time $t = t_f$ to the first instance when a terminating situation is attained. However, when we start to construct candidate optimal trajectories backwards, from points on the target-set into the interior of the game-set, it is more convenient to use *retro-time* τ (sometimes called *time-to-go*) where $\tau = 0$ is associated with the instance of termination. Obviously

$$\tau = t_f - t.$$

The retro-time form of the adjoint equations is:

$$\frac{d}{d\tau} J_x[\mathbf{x}(\tau)] = \frac{\partial}{\partial x} H(\mathbf{x}, \mathbf{J}_x, \overset{*}{\mathbf{u}}, \overset{*}{\mathbf{v}}), \tag{5.37}$$

We shall follow Isaacs' notation and denote differentiation with respect to retro-time τ by the superscript $\circ$,

$$\overset{\circ}{J}_x = H(\mathbf{x}, \mathbf{J}_x, \overset{*}{\mathbf{u}}, \overset{*}{\mathbf{v}}).$$

5.7 Problems

Problem 5.7.1 (The Principle of Optimality)

Consider Lemma 5.4.1,

1. show that if we define

$$J(\mathbf{x}) = \overset{*}{\pi}[\mathbf{x}, t, G(\mathbf{x}(T)), T],$$

where

$$\overset{*}{\pi}[\mathbf{x}, t, G, T] = G[\mathbf{x}(T)] + \int_t^T L[\mathbf{x}(\tau), \overset{*}{\mathbf{u}}(\tau), \overset{*}{\mathbf{v}}(\tau), \tau]d\tau,$$

then an alternative statement of the *principle of optimality* is:

$$\overset{*}{\pi}(\mathbf{x}, t, G, T) = \overset{*}{\pi}[\mathbf{x}, t, \overset{*}{\pi}(\cdot, s, G, T), s].$$

•

Problem 5.7.2 (The Principle of Optimality)

Consider a solution of a game with the general outcome functional of problem 2.8.9.
Consider the optimal trajectory that starts at the time t from the state $\mathbf{x} \in \mathcal{W}$ and terminates on the target-set at the time t_f. Let the state $\check{\mathbf{x}} = \mathbf{x}(\mathbf{s})$ be on the mentioned trajectory at the time s.

1. Prove that:

$$J[\mathbf{x}(t)] = J[\mathbf{x}(s)]e^{-\int_t^s c(x,u,v,\tau)d\tau} + \int_t^s e^{-\int_t^\tau c(x,u,v,z)dz} L(x, u, v, \tau)d\tau. \tag{5.38}$$

•

Problem 5.7.3 (The Principle of Optimality)

Let the *outcome functional* be defined by

$$\overset{*}{\pi}[\mathbf{x}, t, G, T] = \pi(x, t, \overset{*}{\mathbf{u}}(\cdot), \overset{*}{\mathbf{v}}(\cdot), G, T) =$$

$$G[x(T)]e^{-\int_t^T c(x,\overset{*}{u},\overset{*}{v},\tau)d\tau} + \int_t^T e^{-\int_t^\tau c(x,\overset{*}{u},\overset{*}{v},s)ds} L(x, \overset{*}{u}, \overset{*}{v}, \tau)d\tau.$$

1. Show that an alternative statement of the *principle of optimality* is:

$$\overset{*}{\pi}(\mathbf{x}, t, G, T) = \overset{*}{\pi}[\mathbf{x}, t, \overset{*}{\pi}(\cdot, s, G, T), s].$$

•

Problem 5.7.4 (The Principle of Optimality)

consider the case of the outcome functional;

$$\pi(x,t,u(\cdot),v(\cdot),G,T) =$$

$$G[x(T)]e^{-\int_t^T c(x,u,v,\tau)d\tau} + \int_t^T e^{-\int_t^\tau c(x,u,v,s)ds} L(x,u,v,\tau)d\tau.$$

Consider the optimal trajectory that starts at the time t from the state $\mathbf{x} \in \mathcal{W}$ and terminates on the target-set at the time t_f. Let the state $\check{\mathbf{x}} = \mathbf{x}(\mathbf{s})$ be on the mentioned trajectory at the time s.

1. In the context of the *extended solution concept* and for δ-optimal strategies, prove that:

$$J[\mathbf{x}(t)] = J[\mathbf{x}(s)]e^{-\int_t^s c(x,\overset{*}{u}_\delta,\overset{*}{v}_\delta,\tau)d\tau} + \int_t^s e^{-\int_t^\tau c(x,\overset{*}{u}_\delta,\overset{*}{v}_\delta,z)dz} L(x,\overset{*}{u}_\delta,\overset{*}{v}_\delta,\tau)d\tau + |t-s|O(\delta). \tag{5.39}$$

•

Problem 5.7.5 (Game of Kind)

Consider a general case of a *Game of Kind (GOK)*.

1. Argue why the solution of a *game of kind* partitions the *game-set* to regions of equal value separated by semipermeable surfaces (*barriers*).

•

Problem 5.7.6 (Eccentric Homicidal Chauffeur game) *[DS85]*

A fast airplane F has a maximal speed of magnitude 1 and a minimal turn radius of magnitude 1 is pursuing a slower aircraft K. The speed of a slower aircraft K is $\beta < 1$ and it has the capability to commit instantaneous turns. Airplane F controls his motion by choosing a turn rate $\varphi \leq |1|$ and aircraft K controls his motion by choosing a heading angle γ. The equations of motion in a *relative frame* attached to airplane F are:

$$\dot{x} = -y\varphi + \beta \sin\gamma,$$

$$\dot{y} = x\varphi - 1 + \cos\gamma.$$

The target-set $\mathcal{C}$ of the game is an eccentric circle of radius σ. The center of the circle is on the positive y-direction at $y = a$,

$$\mathcal{C} \equiv [(x,y) \mid x^2 + (y-a)^2 = \sigma^2].$$

1. Parametrize points on the target-set by a parameter θ:

$$x = \sigma \sin\theta,$$

$$y = \sigma \cos\theta + a.$$

2. Find the *usable part* and specify θ_{bup}.

Problem 5.7.7 (Eccentric Homicidal Chauffeur game) *[DS85]*

Consider the game of problem 5.7 with a change in the roles of the players. The slower aircraft K is the pursuer wishing to capture the faster airplane F. The reference frame and the equations of motion are the same as in problem 5.7. The target-set is now defined in the coordinate frame of the evader and represent a *vulnerability zone* of the faster airplane with respect to a missile launched by its opponent. The target-set $\mathcal{C}$ of the game is an eccentric circle of radius Ω. The center of the circle is on the positive y-direction at $y = b$,

$$\mathcal{C} \equiv [(x, y) \mid x^2 + (y - b)^2 = \Omega^2].$$

1. Parametrize points on the target-set by a parameter θ:

$$x = \Omega \sin\theta,$$

$$y = \Omega \cos\theta + a.$$

2. Find the *usable part* and specify θ_{bup}.

•

Problem 5.7.8

Discuss the validity of relation (5.18) for a solution in nonadmissible strategies of the type $\overset{*}{\mathbf{u}}=\overset{*}{\mathbf{u}}(\mathbf{v})$. (Player P is basing his strategy on a knowledge of the *current* choice of control by player E).

•

Problem 5.7.9

How would the arguments of the proof of Lemma 5.34 be influenced if we drop the assumption made in section 2.3 that the sets of admissible controls U and V do not depend on the state x.

•

6

Sufficient Conditions

6.1 Introduction

This chapter deals with questions that have both theoretical and practical aspects.
How do we validate a candidate-solution?
How can we be sure that a partial solution is indeed a valid part of the solution before a complete solution is achieved?
Is it possible to get an estimate of the *value function*?

In section 6.2 we discuss a *sufficiency theorem*. If a candidate solution satisfies the conditions of the *sufficiency theorem* it is indeed a valid solution of the game.
Section 6.3 treats the conditions that have to be satisfied by a *partial solution* in order to qualify as a part of a valid solution.
Functions that may provide *estimations* of the *value function* are the subject of section 6.4.

6.2 The Sufficiency Theorem

In general, a candidate solution of an optimization problem that satisfies *local* necessary conditions for optimality cannot be considered as *globally* optimal unless the *complete* set of such candidate-solutions is explored. A candidate-solution of a differential game, satisfying the local necessery conditions for optimality, is no exception with respect to this aspect.

Theorem 6.2.1 (The sufficiency theorem)

If a candidate solution

$$[\mathcal{W}, \mathcal{F}, \overset{*}{U}, \overset{*}{V}, J(\mathbf{x})],$$

satisfies the necessary conditions of the previous chapter for each point of the capture zone $\mathcal{W}$ and if the boundary of the set $\mathcal{W}$ is a semipermeable surface then

$$[\mathcal{W}, \mathcal{F}, \overset{*}{U}, \overset{*}{V}, J(\mathbf{x})],$$

is a *valid solution* of the differential game.

Proof

We shall give a proof for the simple case when the optimal trajectories are

regular. The Hamiltonian function (very much like the *Weierstrass* function in the calculus of variations) plays the key role in the proof.
We have to verify that the *saddle relations*, (see equation 3.1),

$$\pi[\mathbf{x}, \overset{*}{\mathbf{u}}(\cdot), v(\cdot)] \leq J(\mathbf{x}) = \pi[\mathbf{x}, \overset{*}{\mathbf{u}}(\cdot), \overset{*}{\mathbf{v}}(\cdot)] \overset{E}{\preceq} \pi[\mathbf{x}, u(\cdot), \overset{*}{\mathbf{v}}(\cdot)], \tag{6.1}$$

will hold for all the plays $[\mathbf{x}, \overset{*}{\mathbf{u}}(\cdot), v(\cdot)]$, and $[\mathbf{x}, u(\cdot), \overset{*}{\mathbf{v}}(\cdot)]$.

Let us consider a play $[\mathbf{x}, \overset{*}{\mathbf{u}}(\cdot), v(\cdot)]$.
Integrating the Hamiltonian function along the trajectory of the play and recalling lemma 5.2.2 and the definition of the outcome functional (see section 2.6), we obtain:

$$\int_0^{t_f} H(\mathbf{x}, \mathbf{J}_x, \overset{*}{\mathbf{u}}, \mathbf{v}) d\tau = \pi[\mathbf{x}, \overset{*}{\mathbf{u}}(\cdot), \mathbf{v}(\cdot)] - J(\mathbf{x}). \tag{6.2}$$

Similarly if $[\mathbf{x}, \mathbf{u}(\cdot), \overset{*}{\mathbf{v}}(\cdot)]$ is a play that terminates:

$$\int_0^{t_f} H(\mathbf{x}, \mathbf{J}_x, \mathbf{u}, \overset{*}{\mathbf{v}}) d\tau = \pi[\mathbf{x}, \mathbf{u}(\cdot), \overset{*}{\mathbf{v}}(\cdot)] - J(\mathbf{x}). \tag{6.3}$$

Isaacs theorems imply that:

$$H(\mathbf{x}, \mathbf{J}_x, \overset{*}{\mathbf{u}}, \mathbf{v}) \leq 0,$$

$$H(\mathbf{x}, \mathbf{J}_x, \mathbf{u}, \overset{*}{\mathbf{v}}) \geq 0,$$

so evidently relation (6.2) yields:

$$\pi[\mathbf{x}, \overset{*}{\mathbf{u}}(\cdot), \mathbf{v}(\cdot)] - J(\mathbf{x}) \leq 0,$$

and relation (6.3) gives:

$$\pi[\mathbf{x}, \mathbf{u}(\cdot), \overset{*}{\mathbf{v}}(\cdot)] - J(\mathbf{x}) \geq 0.$$

In the case that the play $[\mathbf{x}, \mathbf{u}(\cdot), \overset{*}{\mathbf{v}}(\cdot)]$ happen to stay in the escape region we have that $\pi[\mathbf{x}, \mathbf{u}(\cdot), \overset{*}{\mathbf{v}}(\cdot)] = \Upsilon$ and in view of the preferences assumed for player P we automatically have

$$J(\mathbf{x}) \overset{E}{\preceq} \pi[\mathbf{x}, \mathbf{u}(\cdot), \overset{*}{\mathbf{v}}(\cdot)].$$

•

6.3 Validity of Partial Solutions

The sufficiency theorem (Theorem 6.2.1) implies that we cannot consider a partial construction of the solution of a differential game to be part of a *valid solution* before the construction of the solution is completed, the *capture-set* $\mathcal{W}$ is identified and a candidate value function $J(\mathbf{x})$ is defined over it.
(Equivalently it is required that the *capture-set* $\mathcal{W}$ will be *completely* covered by candidate optimal trajectories) .
This is particularly discouraging in view of the way by which solutions are actually constructed. We gradually cover the game-set with candidate trajectories upon which the necessary conditions applicable to regular points are fulfilled. Often we do not succeed to obtain a complete coverage right away. We must then bare in mind that the fields of candidate optimal trajectories that we constructed cannot even be considered as a part of the solution!
The following lemma states conditions for a case when partial solutions can be considered as valid parts of the complete solution of a differential game.

Lemma 6.3.1 *(***Sufficient Conditions for Partial Solutions***)*

Let a subset $\mathcal{K}$ of the game-set be bounded by a (n-1)-dimensional closed surface $\partial\mathcal{K}$. The surface $\partial\mathcal{K}$ is a composite surface that consists of a subset of the target-set and a smooth surface in the game-set. Let $\mathbf{x}$ belong to $\partial\mathcal{K}$ and let $\mathbf{n}(\mathbf{x})$ be the normal to $\partial\mathcal{K}$ at $\mathbf{x}$ pointing outside $\mathcal{K}$.
If for any $\mathbf{x} \in \partial\mathcal{K}$ that does not belong to the target-set we have that:

$$\max_{\mathbf{v}\in\mathcal{V}} \mathbf{f}(\mathbf{x}, \overset{*}{\mathbf{u}}, \mathbf{v}) \cdot \mathbf{n}(\mathbf{x}) < \mathbf{0},$$

and if the interior of $\mathcal{K}$ is covered with candidate trajectories that satisfy all the necessary conditions then the partial solution inside the set $\mathcal{K}$ is valid.

Proof
The proof involves permeability arguments but beside that it resembles the proof of theorem 6.2.1. The details are left to the reader.

•

6.4 Estimatioms of the Value Function

Many times, for practical reasons or in the course of solving a problem, we wish to have an estimate $Q(\mathbf{x})$ (in the form of a lower or an upper bound) of the value function $J(\mathbf{x})$ in some subset $\mathcal{K}$ of the *game-set* $\mathcal{S}$.

Let us consider the case where the boundary $\partial\mathcal{K}$ of the set $\mathcal{K}$ contains a subset $\mathcal{K}_c$ that belongs to the *target-set* $\mathcal{C}$.

$$\mathcal{K}_c = \partial\mathcal{K} \cap \mathcal{C} \neq \emptyset.$$

Let $\mathbf{n}(\mathbf{x})$ be the normal to the boundary of $\mathcal{K}$ pointing outwards for $\mathbf{x} \in \mathcal{K}_c$ and pointing inwards elsewhere.

Lemma 6.4.1 (The estimating function) [1]

Suppose that a triplet

$$[Q(\mathbf{x}), \tilde{\mathbf{u}}, \tilde{\mathbf{v}}]$$

is given on $\mathcal{K}$, where $(\tilde{\mathbf{u}}, \tilde{\mathbf{v}})$ are *strategies* such that:

1.

$$\begin{array}{lll} H(\mathbf{x}, Q_x, \tilde{\mathbf{u}}, \mathbf{v}) \leq & H(\mathbf{x}, Q_x, \tilde{\mathbf{u}}, \tilde{\mathbf{v}}) \leq & H(\mathbf{x}, Q_x, \mathbf{u}, \tilde{\mathbf{v}}), \\ \forall \mathbf{v} \in \mathcal{V} & & \forall \mathbf{u} \in \mathcal{U} \end{array}$$

$$H(\mathbf{x}, Q_x, \tilde{\mathbf{u}}, \tilde{\mathbf{v}}) \leq 0,$$

2.

$$\begin{array}{ll} \mathbf{f}(\mathbf{x}, \tilde{\mathbf{u}}, \mathbf{v}) \cdot \mathbf{n}(\mathbf{x}) < 0, & \mathbf{f}(\mathbf{x}, \tilde{\mathbf{u}}, \mathbf{v}) \cdot \mathbf{n}(\mathbf{x}) > 0, \\ \forall \mathbf{v} \in \mathcal{V} & \forall \mathbf{v} \in \mathcal{V} \\ \forall \mathbf{x} \in \mathcal{K}_c & \forall \mathbf{x} \text{ elsewhere on } \partial\mathcal{K} \end{array}$$

3.

The plays $[\mathbf{x}, \tilde{\mathbf{u}}(\cdot), \mathbf{v}(\cdot)]$ terminate in finite time for all $\mathbf{x} \in \mathcal{K}$ and for all $\mathbf{v} \in \mathcal{V}$,

4.

$$Q(\mathbf{x}_f) \geq J(\mathbf{x}_f) \quad \text{for} \quad \mathbf{x} \in \mathcal{K}_c,$$

Then:

$$Q(\mathbf{x}) \geq J(\mathbf{x}) \quad \text{for all} \quad \mathbf{x} \in \mathcal{K}.$$

Proof
The pair $(\tilde{\mathbf{u}}, \overset{*}{\mathbf{v}})$ can be considered as a *deviation from optimal* by player E with respect to $Q(\mathbf{x})$:

$$H(x, Q_x, \tilde{u}, \overset{*}{v}) \leq 0,$$

and as a *deviation from optimal* by player P with respect to $J(\mathbf{x})$:

$$H(x, J_x, \tilde{u}, \overset{*}{v}) \geq 0.$$

[1] *A. Friedman* defined a similar set of conditions in a more restrictive context in [Fri72].

By integrating we get:

$$Q[\mathbf{x}(t_f)] - Q[\mathbf{x}(t)] + \int_t^{t_f} L(\mathbf{x}, \tilde{\mathbf{u}}, \overset{*}{\mathbf{v}}) d\tau \leq 0,$$

$$J[\mathbf{x}(t_f)] - J[\mathbf{x}(t)] + \int_t^{t_f} L(\mathbf{x}, \tilde{\mathbf{u}}, \overset{*}{\mathbf{v}}) d\tau \geq 0.$$

Subtracting those relations from each other we obtain:

$$Q[\mathbf{x}(t_f)] - J[\mathbf{x}(t_f)] \leq V[\mathbf{x}(t)] - J[\mathbf{x}(t)],$$

and the lemma follows.

If in (*1.*) we have:

$$H(\mathbf{x}, Q_x, \tilde{\mathbf{u}}, \tilde{\mathbf{v}}) \geq 0,$$

and in (*4.*):

$$Q(\mathbf{x}_f) \leq J(\mathbf{x}_f) \quad \text{for} \quad \mathbf{x} \in \mathcal{K}_c,$$

then a similar argument will give that:

$$Q(\mathbf{x}) \leq J(\mathbf{x}) \quad \text{for all} \quad \mathbf{x} \in \mathcal{K}.$$

•

6.5 Problems

Problem 6.5.1

How should Lemma 6.3.1 be stated if $\partial\mathcal{K}$ is allowed to be non-smooth?

•

Problem 6.5.2

How should we modify Lemma 6.3.1 if for any point $\mathbf{x}$ on $\partial\mathcal{K}$ that does not belong to the target-set we have:

$$\max_{\mathbf{v}\in\mathcal{V}} \mathbf{f}[\mathbf{x}, \overset{*}{\mathbf{u}}, \mathbf{v}] \cdot \mathbf{n}(\mathbf{x}) \leq 0.$$

•

7

Construction of Regular Candidate Solutions

7.1 Introduction

The theory of differential games accommodates a veriety of phenomena that are necessary in order to model conflicts in the real world. Not all the phenomena show up in every problem and even if they do they govern specific *singular* manifolds while in most of the game set we have the simple regular case.
For this reason a systematic approach for solving differential games relies on two kinds of procedures. A *regular* procedure, by which we construct regular candidate partial solutions, and a *singular* procedure that provides a mode of joining candidate regular partial solutions across singular surfaces.

In this chapter we begin to discuss such a systematic approach and we describe how the first attempt to construct a regular partial solution of a problem is done. We shall use examples in order to clarify the subject.
The algorithm for the construction of candidate optimal trajectories[1] is based on the idea of going backwards in time from points on the target set where optimal plays end (this is indeed similar to the dynamic programming approach).
Section 7.2 describes the *regular* procedure. It discusses the steps of the procedure and its fundamental properties.
In section 7.3 we check the validity of the candidate solutions for all the examples. Some examples have a valid solution at this stage while the other examples demonstrate that further treatment that involves procedures for constructing singular parts of the solutions is required.
The chapter also includes a section on *linear quadratic differential games*. As a matter of fact the solution to an important class of this family of games is more easily obtained by solving a *Riccati* equation (resembling *linear quadratic control problems*) rather than by using the methods of Isaacs that are the subject of this book.
The basic theory is covered to the degree that suffices to correlate it to the theory of $\mathcal{H}^\infty$ optimal control, (a subject that has gained lots of attention

[1] Mathematicians relates the solution of a differential game to the solution of a parabolic partial differential equation (the Isaac's equations) and regards the optimal trajectories as *characteristics* of the solution.

in the recent years).

7.2 The Regular Procedure

7.2.1 Partitioning the Target-Set

We recall from section 5.2.1 that optimal trajectories that start at points $\mathbf{x}$ that belong to the *capture-set* $\mathcal{W}$ terminate at points that belong to the *usable part* of the *target-set*[2] .
When we solve a differential game the first step is therefore to identify the *usable part* (*UP*), the *nonusable part* (*NUP*) and the *boundary of the usable part* (*BUP*) of the target-set.

We first evaluate $D(\mathbf{x})$,

$$D(\mathbf{x}) = \min_{\mathbf{u}\in\mathcal{U}} \max_{\mathbf{v}\in\mathcal{V}} \mathbf{f}(\mathbf{x},\mathbf{u},\mathbf{v})\cdot\mathbf{n}(\mathbf{x}),$$

where $\mathbf{n}(\mathbf{x})$ is a vector normal to the target-set at $\mathbf{x}$ pointing into the game-set.

The desired partitioning of the target-set is given by:

$$\begin{aligned} \mathrm{UP} &\equiv \{\mathbf{x} \mid D(\mathbf{x}) < 0\}, \\ \mathrm{NUP} &\equiv \{\mathbf{x} \mid D(\mathbf{x}) > 0\}, \qquad (7.1) \\ \mathrm{BUP} &\equiv \{\mathbf{x} \mid D(\mathbf{x}) = 0\}. \end{aligned}$$

7.2.2 Candidate Optimal Control Laws

The second step is to use *Isaac's equations* in order to find candidate optimal control laws. We solve two local optimization problems:

$$\max_{\mathbf{v}\in\mathcal{V}} H(\mathbf{x},\mathbf{J}_x,\overset{*}{\mathbf{u}},\mathbf{v}) = H(\mathbf{x},\mathbf{J}_x,\overset{*}{\mathbf{u}},\overset{*}{\mathbf{v}}), \qquad (7.2)$$

$$\min_{\mathbf{u}\in\mathcal{U}} H(\mathbf{x},\mathbf{J}_x,\mathbf{u},\overset{*}{\mathbf{v}}) = H(\mathbf{x},\mathbf{J}_x,\overset{*}{\mathbf{u}},\overset{*}{\mathbf{v}}), \qquad (7.3)$$

and we get expressions for $\overset{*}{\mathbf{u}}$ and $\overset{*}{\mathbf{v}}$ that depend on $\mathbf{x}$ and $\mathbf{J}_x(\mathbf{x})$.
$\overset{*}{\mathbf{u}}$ and $\overset{*}{\mathbf{v}}$ are *candidate* closed-loop optimal control laws for player P and player E respectively.

---

[2]The opposite is not true, not all the points of the *UP* are necessarily end points of optimal trajectories.

Notice that when we use equation (7.2) and equation (7.3) we have first to assume that $\mathbf{J}_x$ is continuous at $\mathbf{x}$ (which is the same as assuming that $\mathbf{x}$ belongs to a *regular* part of an optimal trajectory). The validity of this assumption should be checked as we proceed!

7.2.3 Retro-Integration of the Adjoint Equations

We want to construct candidate optimal trajectories by retro-integrating

$$\overset{\circ}{\mathbf{x}} = -\mathbf{f}(\mathbf{x}, \overset{*}{\mathbf{u}}, \overset{*}{\mathbf{v}}) \quad , \ \mathbf{x}(0) = x_0,$$

but in subsection 7.2.2 we saw that in order to obtain the current values of the expressions for $\overset{*}{\mathbf{u}}$ and $\overset{*}{\mathbf{v}}$ we need to know the current values of the *adjoint variables* $\mathbf{J}_x(\mathbf{x})$ too. For this reason we have to *retro-integrate* the *adjoint equations*,

$$\overset{\circ}{\mathbf{J}}_x = \frac{\partial H(\mathbf{x}, \mathbf{J}_x, \overset{*}{\mathbf{u}}, \overset{*}{\mathbf{v}})}{\partial x} \quad , \ \mathbf{J}_x(\mathbf{x}(0)) = \mathbf{J}_x(\mathbf{x}_0), \tag{7.4}$$

simultaneously.

The initial values $\mathbf{J}_x(\mathbf{x}_0)$ are not given explicitly and we have to figure them out. From Corollary 5.2.3 we have $(n-1)$ independent relations involving the n components of $\nabla J(\mathbf{x}_0)$:

$$\nabla J(\mathbf{x}_0) \cdot \mathbf{t}_i = \nabla G(\mathbf{x}_0) \cdot \mathbf{t}_i, \tag{7.5}$$

where $\mathbf{t}_i$ is any vector that is tangent to the target-set at $\mathbf{x}_0$.

Notice that relations (7.5) do not involve the component of $\nabla J(\mathbf{x}_0)$ in a the direction of the normal to the target-set at $\mathbf{x}_0$.
We need another relation between the n components of $\nabla J(\mathbf{x}_0)$ that will be linearly independent of relations (7.5). We may obtain this additional relation by specializing ME2 (equation 5.11) to $\mathbf{x}_0$,

$$H(\mathbf{x}_0, \mathbf{J}_x, \overset{*}{\mathbf{u}}, \overset{*}{\mathbf{v}}) = 0. \tag{7.6}$$

At $\mathbf{x}_0$ *ME2* is indeed independent of relations (7.5) because of the requirement that $[\mathbf{x}_0, \overset{*}{\mathbf{u}}, \overset{*}{\mathbf{v}}]$ will be a *termination situation*:

$$\mathbf{f}(\mathbf{x}_0, \overset{*}{\mathbf{u}}, \overset{*}{\mathbf{v}}) \cdot \mathbf{n}(\mathbf{x}_0) < 0.$$

$\mathbf{f}(\mathbf{x}_0, \overset{*}{\mathbf{u}}, \overset{*}{\mathbf{v}})$ is hence *nontangent* to the target-set at $\mathbf{x}_0$ and equation (7.6) involves the component of $\mathbf{J}_x(\mathbf{x}_0)$ in the direction of the normal to the target-set at $\mathbf{x}_0$.

Remarks

1. In order to retro-integrate the adjoint equations we needed to use the information contained in ME2 only at the initiation state $\mathbf{x}_0$. We know that ME2 is valid along *any* regular part of an optimal trajectory. ME2 is indeed an integral relation[3] between the components of $\mathbf{J}_x(\mathbf{x})$. In the course of the computation, we can use this redundant information[4] to replace the integration of one of the n adjoint equations by ME2[5].

2. When we start to construct a candidate optimal trajectory from a point $\mathbf{x}_0$ on the *usable part* we assume that an optimal trajectory terminates there. It is important to remember that this assumption may not always be valid and we may find, at a later stage of the construction, that no optimal trajectory terminates at this point in spite of the fact that it belongs to the *usable part*.

3. The reader should be aware that the adjoint equations are valid only along *regular* parts of an optimal trajectory. It is *wrong* to infer that a part of a candidate optimal trajectory is regular just from the fact that it was derived by our algorithm, before ascertaining that the construction provided a *complete solution*.

7.2.4 Properties of the Manifolds of Candidate Optimal Trajectories

While we construct candidate optimal trajectories we substitute values of the *adjoint variables* (obtained from the integration of the adjoint equations) in the *Isaacs' equations* assuming they are components of the gradient of the candidate value function of the game. Let us show that this is indeed the case.

We consider the case when the state of the game is a n-dimensional vector x[6] , the *target-set* is a $(n-1)$-dimensional set and it is parametrized by a $(n-1)$ vector σ of parameters,

$$x = x(\sigma).$$

Lemma 7.2.1 (The Manifold Lemma)

---

[3] The adjoint equations are differential relations between the components of $\mathbf{J}_x(\mathbf{x})$.

[4] Recall that we derived the adjoint equations through the validity of ME2 on neighboring optimal paths.

[5] This resembles what is often done with the *energy equations* in classical mechanics.

[6] In this subsection vector quantities are not printed in bold print.

If a given region $\mathcal{R}$ is completely covered by candidate optimal trajectories emanating from points of the *UP* constructed by *retro-integration* of

$$\frac{\partial \overset{*}{x}(\tau,\sigma)}{\partial \tau} = -f(\overset{*}{x},\overset{*}{u},\overset{*}{v}) \ , \quad \overset{*}{x}(0,\sigma) = x(\sigma), \tag{7.7}$$

$$\frac{\partial \lambda[\overset{*}{x}(\tau,\sigma)]}{\partial \tau} = \frac{\partial}{\partial \overset{*}{x}} H(\overset{*}{x},\lambda,\overset{*}{u},\overset{*}{v}) \ , \quad \lambda[\overset{*}{x}(0,\sigma)] = G_x[x(\sigma)], \tag{7.8}$$

where the optimal controls $\overset{*}{u}$ and $\overset{*}{v}$ satisfy Isaacs' equations. then the *candidate value function* is

$$W[\overset{*}{x}(\tau,\sigma)] = G[x(\sigma)] - \int_{\tau}^{0} L[\overset{*}{x}(\alpha,\sigma),\overset{*}{u},\overset{*}{v}]d\alpha, \tag{7.9}$$

and,

$$\lambda(\overset{*}{x}(\tau,\sigma)) = \frac{dW[\overset{*}{x}(\tau,\sigma)]}{d\overset{*}{x}}. \tag{7.10}$$

Proof[7].

Let an arbitrary curve C in the region $\mathcal{R}$ be defined by the following C^2 functions:

$$\tau = \tau_0(s), \qquad x = X_0(s), \qquad 0 \le s \le 1. \tag{7.11}$$

The end points of the members of the manifold of candidate optimal trajectories that pass through the curve C define a curve $\sigma = \sigma(s)$ in the *target-set*. A member trajectory of this manifold can hence be described by:

$$x = \overset{*}{x}(\tau,\sigma(s)).$$

The starting point $X_0(s)$ of a member trajectory lies on the curve C and its end point $X_f(s)$ belongs to the curve $\sigma = \sigma(s)$ on the target-set:

$$\begin{aligned} X_0(s) &= \overset{*}{x}(\tau_0(s),\sigma(s)), \\ X_f(s) &= \overset{*}{x}(0,\sigma(s)), \end{aligned} \tag{7.12}$$

and the candidate value function that correspond to points of the curve C is:

$$W[X_0(s)] = G[X_f(s)] - \int_{\tau_0(s)}^{0} L[\overset{*}{x}(\alpha,\sigma(s)),\overset{*}{u},\overset{*}{v}]d\alpha. \tag{7.13}$$

[7] We give a slightly simplified version of the proof given by *L. D. Berkovitz* in [Ber64].

As we move along C from $X_0(s)$ to $X_0(s+\Delta s)$ we have[8]

$$\frac{dW}{ds} = G_x[X_f(s)]\frac{dX_f}{ds} + \\ - \int_{\tau_0(s)}^{0}(\frac{\partial L}{\partial \overset{*}{x}}\overset{*}{x}_\sigma \frac{d\sigma}{ds})d\alpha + L[X_0(s), \overset{*}{u}(X_0(s)), \overset{*}{v}(X_0(s))]\frac{d\tau_0(s)}{ds}. \tag{7.14}$$

Notice that in equation (7.14) and in the following arguments we denote:

$$\frac{\partial L}{\partial \overset{*}{x}} = L_{\overset{*}{x}} + L_u \overset{*}{u}_{\overset{*}{x}} + L_v \overset{*}{v}_{\overset{*}{x}}, \\ \frac{\partial f}{\partial \overset{*}{x}} = f_{\overset{*}{x}} + f_u \overset{*}{u}_{\overset{*}{x}} + f_v \overset{*}{v}_{\overset{*}{x}}. \tag{7.15}$$

From the adjoint equation we obtain that[9]:

$$\frac{\partial L}{\partial \overset{*}{x}}\overset{*}{x}_\sigma = \lambda_\tau \overset{*}{x}_\sigma + \lambda \frac{\partial f}{\partial \overset{*}{x}}\overset{*}{x}_\sigma = (\lambda \overset{*}{x}_\sigma)_\tau, \tag{7.16}$$

so

$$\int_{\tau_0(s)}^{0}(\frac{\partial L}{\partial \overset{*}{x}}\overset{*}{x}_\sigma \frac{d\sigma}{ds})d\alpha = \frac{d\sigma}{ds}\int_{\tau_0(s)}^{0}(\lambda \overset{*}{x}_\sigma)_\alpha d\alpha = \tag{7.17}$$

$$= \frac{d\sigma}{ds}[\lambda(X_f(s))\overset{*}{x}_\sigma(0, X_f(s)) - \lambda(X_0(s))\overset{*}{x}_\sigma(\tau_0(s), X_0(s))].$$

But recalling (7.12):

$$\frac{dX_0(s)}{ds} = -\overset{*}{f}(\tau_0(s), X_0(s))\frac{d\tau_0}{ds} + \overset{*}{x}_\sigma(\tau_0(s), X_0(s))\frac{d\sigma(s)}{ds}, \tag{7.18}$$

and

$$\frac{dX_f(s)}{ds} = \overset{*}{x}_\sigma(0, X_f(s))\frac{d\sigma(s)}{ds}. \tag{7.19}$$

Substituting from equations (7.18) and (7.19) into equation (7.17) and then into equation (7.14) and rearranging[10] we obtain:

$$\frac{dW}{ds} = [\lambda(0, X_f(s)) - G_x(X_f(s)]\frac{dX_f}{ds}(s) - \overset{*}{H}(\tau_0(s), X_0(s))\frac{d\tau_0(s)}{ds} + \\ + \lambda(\tau_0, X_0(s))\frac{dX_0}{ds}(s). \tag{7.20}$$

---

[8] For simplicity we shall denote a matrix $\frac{\partial a_i}{\partial b_j}$ as a_b. We denote $\frac{\partial \overset{*}{x}(\tau,\sigma)}{\partial \sigma}$ as $\overset{*}{x}_\sigma$ in order to distinguish it from $\frac{\partial x(\sigma)}{\partial \sigma}$.

[9] We use the fact that: $\overset{*}{x}_{\tau\sigma} = \frac{\partial f}{\partial \overset{*}{x}}\overset{*}{x}_\sigma = \overset{*}{x}_{\sigma\tau}$.

[10] We denote:

$$H[\overset{*}{x}(\tau_0(s), X_0(s)), \lambda, \overset{*}{u}, \overset{*}{v}] = \overset{*}{H}(\tau_0(s), X_0(s)).$$

Recall that on the UP,

$$\lambda(0, X_f(s)) - G_x(X_f(s) = 0,$$

and that the construction of a candidate optimal trajectory provides

$$\overset{*}{H}(\tau_0(s), X_0(s)) = 0.$$

Equation (7.20) now reduces to:

$$\frac{dW(X_0(s))}{ds} = \lambda(\tau_0, X_0(s))\frac{dX_0}{ds}(s). \tag{7.21}$$

The curve C was chosen arbitrarily so

$$dW = \lambda dX, \tag{7.22}$$

holds for arbitrary dX.

•

Because of the *Manifold Lemma* we have that if the regular construction procedure succeeds to fill the *game-set* with candidate optimal trajectories or to define a *capture-set* $\mathcal{W}$ it also satisfies the sufficient condition of Lemma 6.2.1.

7.3 Examples

Let us examine some examples.

Example 7.3.1 (The preview example)

In section 1.2 we solved a preview example using heuristic arguments. Let us resolve it with the *regular procedure*.
We first cast the problem in the format of a game model:

The game-set:

$$\mathcal{S} \equiv \{(x, y) \mid y \geq 0\}.$$

The target-set:

$$\mathcal{C} \equiv \{(x, y) \mid y = 0\}.$$

The equations of motion:

$$\dot{x} = Av + B\sin u \ , \quad |v| \leq 1,$$

$$\dot{y} = -1 + B\cos u.$$

A and B are constants and $0 < B < A < 1$. Player P chooses u and player E chooses v.

The outcome functional:

$$G(x,0) = x.$$

Solution:

Partitioning of the target-set
The normal to the target-set at a point $(x,0)$ is $\mathbf{n}(x,0) = \hat{\mathbf{e}}_y$, where $\hat{\mathbf{e}}_y$ is a unit vector in the y-direction. We have:

$$\mathbf{f}(x,y,u,v) \cdot \mathbf{n}(x,0) = -1 + B\cos u < 0,$$

so evidently

$$\mathrm{UP} \equiv \mathcal{C},$$

$$\mathrm{NUP} \equiv \emptyset,$$

where $\emptyset$ denotes an *empty set*.

Candidate optimal control laws
The equations of motion are *separable* so we can express the Isaacs equation ME1 as:

$$\min_u \max_{|v|\leq 1} \{(Av + B\sin u)J_x + (-1 + B\cos u)J_y\}.$$

We can derive the candidate optimal contol laws almost by inspection. The optimal control for player E is:

$$\begin{aligned} \overset{*}{v}(x,y) &= \operatorname{sign} J_x(x,y)\ , & \text{for } J_x(x,y) \neq 0, \\ \overset{*}{v}(x,y) &= \text{ any admissible } v\ , & \text{for } J_x(x,y) = 0. \end{aligned} \tag{7.23}$$

The *sign* function that appears in equation (7.23) is typical to cases where a component of the control vector is *linear*. The term $J_x(x,y)$ in equation (7.23) plays the role of a *switch function* for the optimal control $\overset{*}{v}(x,y)$.

The optimal control for player P is:

$$(\sin \overset{*}{u}(x,y), \cos \overset{*}{u}(x,y)) \parallel (-J_x(x,y), -J_y(x,y)) \tag{7.24}$$

Equation (7.24), the optimal control law for $\overset{*}{u}(x,y)$, means that the vector with the components $(\sin \overset{*}{u}(x,y), \cos \overset{*}{u}(x,y))$ is collinear (*aligned*) with the vector $(-J_x(x,y), -J_y(x,y))$.
This is typical for the solution of optimization problems that can be regarded as problems of maximization (or minimization) of the *inner product*

of vectors.

The adjoint equations
We have:

$$\overset{\circ}{J}_x\ [x(\tau), y(\tau)] = 0,$$
$$\overset{\circ}{J}_y\ [x(\tau), y(\tau)] = 0. \tag{7.25}$$

We recall that at points $\mathbf{x}$ on the usable-part, where optimal trajectories terminate, we have that for any vector $\mathbf{t}$ tangent to the target-set:

$$\nabla J(\mathbf{x}) \cdot \mathbf{t} = \nabla G(\mathbf{x}) \cdot \mathbf{t}. \tag{7.26}$$

Here $\mathbf{t} = \hat{\mathbf{e}}_x$. ($\hat{\mathbf{e}}_x$ is a unit vector in the x-direction).
We obtain:

$$J_x(x, 0) = 1. \tag{7.27}$$

In order to find $J_y(x, o)$ we have to utilize ME2. Let us substitute the optimal controls in ME1 to obtain ME2.

$$A|J_x(x,y)| - J_y(x,y) - B\sqrt{J_x^2(x,y) + J_y^2(x,y)} = 0. \tag{7.28}$$

At $\tau = 0$, we get, after substituting from equation (7.27),

$$A - J_y(x,0) - B\sqrt{1 + J_y^2(x,0)} = 0, \tag{7.29}$$

and we can solve for $J_y(x,0)$[11].

The candidate optimal trajectories
From equation (7.25), (7.24) and (7.23) we can deduce that $J_x(x,y)$ and $J_y(x,y)$ as well as $\overset{*}{u}$ and $\overset{*}{v}$ remain constant along each optimal trajectory. It follows that the optimal trajectories are parallel straight lines.
The field of candidate optimal trajectories covers the game-set completely so the sufficiency conditions are satisfied and it is a *valid solution* of the game.

•

Example 7.3.2

This is also a very simple problem. We use it to show that the two types of *order of preference* (*type P* where player P considers nontermination as the worst result, and *type E* where player E despises nontermination most) induce *different* solutions!

---

[11] Only one of the solutions of this quadratic equation is correct. The choice is determined by meeting the termination requirement at $(x, 0)$.

The game-set:

$$\mathcal{S} \equiv \{(x, y) \mid y > 0 \ \ , x + y - 1 \leq 0\}.$$

The target-set:

$$\mathcal{C} \equiv \{(x, y) \mid y > 0 \ \ , x + y - 1 = 0\}.$$

The equations of motion:

$$\dot{x} = 1 + v \ , \qquad |v| \leq 1,$$

$$\dot{y} = 1 + u \ , \qquad |u| \leq 1.$$

Player P chooses u and player E chooses v.

The outcome functional:

$$G(x, y) = y^2 - x^2 \ , \qquad L(x, y) = 0.$$

This is a *terminal cost game*.
We shall distinguish between two cases regarding the *order of preference*.

1. Player P considers non-termination worse than any numerical result.

2. Player E considers non-termination worse than any numerical result.

Solution of case 1.

The partitioning of the target-set
The normal to the target-set at a point (x, y) pointing into the game-set is

$$\mathbf{n}(x, y) = -\hat{\mathbf{e}}_x - \hat{\mathbf{e}}_y \, ,$$

where $\hat{\mathbf{e}}_x$ and $\hat{\mathbf{e}}_y$ are unit vectors in the x-direction and the y-direction respectively. The termination condition is:

$$\min_{|u| \leq 1} \max_{|v| \leq 1} \mathbf{f}(x, y, u, v) \cdot \mathbf{n}(x, y) < 0.$$

It is easy to see that if player P chooses $u \neq -1$ he can guarantee termination. So:

$$\text{UP} \equiv \mathcal{C},$$

$$\text{NUP} \equiv \emptyset.$$

Candidate optimal control laws
The equations of motion are *separable* so we can express the Isaacs equation ME1 at *regular* points as:

$$\min_{|u| \leq 1} \max_{|v| \leq 1} \{(1 + v) J_x(x, y) + (1 + u) J_y(x, y)\}. \tag{7.30}$$

The candidate optimal control laws for player P are:

$$\overset{*}{u} = \text{-sign} J_y(x,y) \;, \qquad \text{for } J_y(x,y) \neq 0,$$
$$\overset{*}{u} = \text{ any admissible } u \;, \quad \text{for } J_y(x,y) = 0. \tag{7.31}$$

The candidate optimal control laws for player E are:

$$\overset{*}{v} = \text{sign} J_x(x,y) \;, \qquad \text{for } J_x(x,y) \neq 0,$$
$$\overset{*}{v} = \text{ any admissible } v \;, \quad \text{for } J_x(x,y) = 0. \tag{7.32}$$

The adjoint equations

$$\overset{\circ}{J_x} (x,y) = 0,$$
$$\overset{\circ}{J_y} (x,y) = 0. \tag{7.33}$$

Requiring that on the *usable-part*

$$\nabla J(x,y) \cdot \mathbf{t} = \nabla G(x,y) \cdot \mathbf{t},$$

and recalling that the tangent to the target-set at (x,y) is

$$\mathbf{t} = -\hat{\mathbf{e}}_x + \hat{\mathbf{e}}_y,$$

We obtain:

$$J_x(x,y) - J_y(x,y) = -2. \tag{7.34}$$

At termination, the requirement $\overset{*}{u} \neq -1$ implies that $J_y(x,y) \leq 0$.
The possibility that at the usable-part $J_y(x,y) < 0$ is excluded because it cannot satisfy both ME2 and equation (7.34) and we get:

$$J_x(x,y) = -2 \;, \quad J_y(x,y) = 0. \tag{7.35}$$

The optimal trajectories
It is easy to see that the optimal trajectories are parallel *horizontal* lines and that they cover the whole game-set.
The value function of the game is:

$$J(x,y) = 1 - 2x. \tag{7.36}$$

Solution of case 2.

This is the case where player E prefers termination over any numerical result of a play.

The partitioning of the target-set
The termination condition for this case is:

$$\min_{|v|\leq 1}\max_{|u|\leq 1}\mathbf{f}(x,y,u,v)\cdot\mathbf{n}(x,y)<0.$$

It is easy to see that if player E chooses $v \neq -1$ he can guarantee termination. So:

$$\mathrm{UP} \equiv \mathcal{C},$$

$$\mathrm{NUP} \equiv \emptyset.$$

Candidate optimal control laws
The candidate optimal control laws at a point (x,y) (assumed to be on a regular part of an optimal trajectory!) are the same as in case 1.

The adjoint equations
As in case a. we have that the components of $\nabla J(x,y)$ on the target-set have to satisfy:

$$J_x(x,y) - J_y(x,y) = -2. \tag{7.37}$$

At termination, the requirement $\overset{*}{v} \neq -1$ implies that $J_x(x,y) \geq 0$.
The possibility that at the usable-part $J_x(x,y) > 0$ is excluded because it cannot satisfy both ME2 and equation (7.37) and we get:

$$J_x(x,y) = 0\ , \quad J_y(x,y) = 2. \tag{7.38}$$

The optimal trajectories
It is easy to see that the optimal trajectories are parallel *vertical* lines and that they cover the whole game-set.
The value function of the game is:

$$J(x,y) = 2y - 1. \tag{7.39}$$

∘

Example 7.3.3

This example is a reduced order model for the problem of proportional navigation studied by *S. Gutman* and *G. Leitmann* [GL76].

The game-set:

$$\mathcal{S} \equiv \{(x,\tau) \mid \tau \geq 0\}.$$

The target-set:

$$\mathcal{C} \equiv \{(x,\tau) \mid \tau = 0\}.$$

The equations of motion:

$$\dot{x} = (u+v)\tau \ , \quad |u| \le U_m \ , \quad |v| \le V_m,$$

$$\dot{\tau} = -1.$$

Player P chooses u and player E chooses v.

The outcome functional:

$$G(x,\tau) = 0 \ , \quad L(x,\tau) = x\tau(u+v).$$

This is a *running cost game*.

Solution

Partitioning of the target-set

The normal to the target-set at a point $(x,0)$ is $\mathbf{n}(x,0) = \hat{\mathbf{e}}_\tau$, where $\hat{\mathbf{e}}_\tau$ is a unit vector in the τ-direction. We have:

$$\mathbf{f}(x,\tau,u,v) \cdot \mathbf{n}(x,0) = -1.$$

So evidently

$$\text{UP} \equiv \mathcal{C},$$

$$\text{NUP} \equiv \emptyset.$$

Candidate optimal control laws

The equations of motion are *separable* so we can express the Isaacs equation ME1 as:

$$\min_{|u| \le U_m} \max_{|v| \le V_m} \{(u+v)\tau J_x(x,\tau) - J_\tau(x,\tau) + x(u+v)\tau\}.$$

The optimal control for player P is:

$$\begin{aligned} \overset{*}{u} &= -U_m \text{sign}[J_x(x,\tau)+x]\tau \ , && \text{for } \tau[J_x(x,\tau)+x] \neq 0, \\ \overset{*}{u} &= \text{ any admissible } u \ , && \text{for } \tau[J_x(x,\tau)+x] = 0. \end{aligned} \tag{7.40}$$

The optimal control for player E is:

$$\begin{aligned} \overset{*}{v} &= V_m \text{sign}[J_x(x,\tau)+x]\tau \ , && \text{for } \tau[J_x(x,\tau)+x] \neq 0, \\ \overset{*}{v} &= \text{ any admissible } v \ , && \text{for } \tau[J_x(x,\tau)+x] = 0. \end{aligned} \tag{7.41}$$

The function $A(x,\tau) = \tau[J_x(x,\tau)+x]$ is the *switch function* for both controls for $\tau > 0$.

The adjoint equations

We have:

$$\overset{\circ}{J}_x\ [(x(\tau),\tau] = (\overset{*}{u} + \overset{*}{v})\tau,$$
$$\overset{\circ}{J}_\tau\ [x(\tau),\tau] = (\overset{*}{u} + \overset{*}{v})[J_x(x,\tau) + x]. \tag{7.42}$$

It is of benefit to give also an expression for $\overset{\circ}{A}\ (x,\tau)$:

$$\overset{\circ}{A}\ (x,\tau) = 0. \tag{7.43}$$

The tangent **t** to the target-set is given by:

$$\mathbf{t} = \hat{\mathbf{e}}_x,$$

where $\hat{\mathbf{e}}_x$ is a unit vector in the x-direction). From the requirement that for points (x,τ) of the usable-part where optimal trajectories terminate we have:

$$\nabla J(x,\tau)\cdot\mathbf{t} = \nabla G(x,\tau)\cdot\mathbf{t}. \tag{7.44}$$

We obtain:

$$J_x(x,0) = 0.$$

Let us substitute the optimal controls in ME1 to obtain ME2.

$$(V_m - U_m)\tau|J_x(x,\tau)| - J_\tau(x,\tau) = 0. \tag{7.45}$$

At $\tau = 0$ we shall have:

$$J_\tau(x,\tau) = 0. \tag{7.46}$$

The optimal trajectories

On the target-set where $\tau = 0$ the switch function for both controls vanishes. But adjacent to the target-set (and elsewhere along an optimal trajectory) we have:

$$A(x,\tau) = x(0), \tag{7.47}$$

and the candidate optimal trajectories are:

$$x(\tau) - x(0) = \frac{1}{2}(V_m - U_m)\tau^2 \mathrm{sign}x(0). \tag{7.48}$$

In Figure 7.1 we see the fields of candidate optimal trajectories for the case when $V_m < U_m$ and for the case when $V_m > U_m$. We have to distinguish between two cases.

1. $V_m < U_m$
   The value function is:

$$J(x,\tau) = \tfrac{1}{2}\ [x - \tfrac{1}{2}\ (U_m - V_m)\tau^2]^2 - \tfrac{1}{2}\ x^2,$$
$$\text{for } |x| > \tfrac{1}{2}\ (U_m - V_m)\tau^2.$$

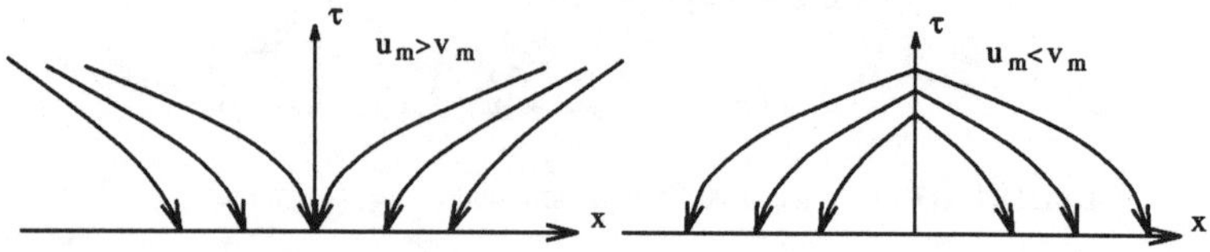

FIGURE 7.1. Optimal trajectories for Example 7.3.3

For $|x| < \frac{1}{2}\,(U_m - V_m)\tau^2$, there is a *void* in the game-set which is not covered by candidate optimal trajectories.
This demonstrates a phenomenon, common to most nontrivial problems, that the field of candidate optimal trajectories constructed by the *regular procedure* backward from the usable-part neither fill the game-set nor define a *capture-set.*
In order to proceed with the solution we must assume the existence of a *singular surface* on which assumptions that where essential for the *regular procedure* are not valid.

In our case we assume that in the *void* the value does not depend on τ:

$$J(x,\tau) = -\tfrac{1}{2}\,x^2\,, \quad \text{for } |x| \leq \tfrac{1}{2}\,(U_m - V_m)\tau^2. \tag{7.49}$$

In this region we have that $A(x,\tau) = 0$ so that any choice of controls for both players is equally optimal.
Recall that when we derived the *adjoint equations* (equation (5.33)) we assumed that the second derivatives of the value function are continuous along the optimal trajectories. It is easy to see that both $J(x,\tau)$, $J_x(x,\tau)$ and $J_\tau(x,\tau)$ are continuous across the boundaries of the void but the second derivatives $J_{xx}(x,\tau)$ and $J_{\tau\tau}(x,\tau)$ are not!

2. $V_m > U_m$
The value function is

$$J(x,\tau) = \frac{1}{2}\,[x - \frac{1}{2}\,(U_m - V_m)\tau^2]^2 - \frac{1}{2}\,x^2.$$

Here the two symmetric families of candidate optimal trajectories intersect.

This demonstrates another common phenomenon that candidate optimal trajectories constructed by the *regular procedure* backward from the usable-part intersect causing ambiguity in the value function and its derivatives. In order to proceed with the solution we must again assume the existence of a *singular surface* on which assumptions that where essential for the *regular procedure* are not valid.

Here we shall assume that the line $x = 0$ is a locus where the evader can choose which way to evade but both alternatives yield the same optimal outcome. Across the line $x = 0$ the function $\nabla J(x, \tau)$ is discontinuous. We shall discuss this kind of *singularity* in more detail in chapter 11.

•

Example 7.3.4 (The Lady In The Lake) *[Isa75]*

The problem is about a *lady* E who swims (with speed $\beta < 1$) in a circular pond (with a radius of magnitude 1). A *lusty man* P runs along the circumference of the pond wishing to take the closest picture of the lady as she gets out...[12].

The game model:

We formulate the game model of this problem in a *relative* polar reference frame. The origin will be fixed at the center of the pond and the $(\theta = 0)$-axis will be attached to P. (See Figure 7.2).

The game-set:

$$\mathcal{S} \equiv \{(r, \theta) \mid r \leq 1\}.$$

The target-set:

$$\mathcal{C} \equiv \{(r, \theta) \mid r = 1\}.$$

The equations of motion:

$$\dot{r} = \beta \cos \psi,$$

$$\dot{\theta} = \tfrac{1}{r}\beta \sin \psi - \varphi\,, \quad |\varphi| \leq 1\,, \quad -\pi \leq \theta < \pi.$$

---

[12] We assume that once the lady is outside the pond she can run faster. If $\frac{1}{\beta}\sqrt{1-\beta^2} < \frac{3}{2}\pi + \sin^{-1}\frac{1}{\beta}$ there is no point-capture and only a picture can be taken...

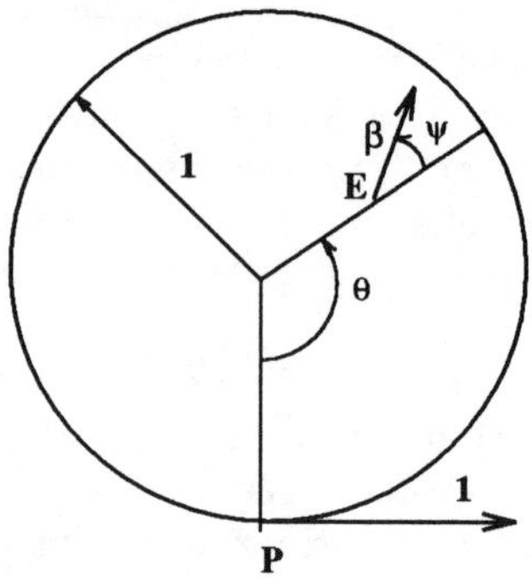

FIGURE 7.2. The Lady in the Lake

Player P chooses φ and player E chooses ψ.

The outcome functional:

$$G(r,\theta) = |\theta|\ , \quad L(r,\theta,\varphi,\psi) = 0.$$

This is a game with a terminal cost. The evader, player E, is assumed to prefer termination over any other result of a play.

Solution:

It is possible to solve the problem by using geometric arguments but it is more instructive to solve it by the mechanism of the theory of differential games.

The partitioning of the target-set
The normal to the target-set at a point $(1,\theta)$ is $\mathbf{n}(1,\theta) = -\hat{\mathbf{e}}_r$ where $\hat{\mathbf{e}}_r$ is a unit vector in the r-direction. We have:

$$\mathbf{f}(r,\theta,\varphi,\psi)\cdot\mathbf{n}(1,\theta) = -\beta\cos\psi.$$

Evidently:

$$\mathrm{UP} \equiv \mathcal{C},$$

$$\mathrm{NUP} \equiv \emptyset.$$

Candidate optimal control laws
The equations of motion are *separable* so for regular parts of candidate optimal trajectories we can express the Isaacs equation ME1 as:

$$\min_{\varphi\leq 1}\max_{\psi}\{(\beta\cos\psi)J_r + [-\varphi + \frac{\beta\sin\psi}{r}]J_\theta\}. \tag{7.50}$$

The candidate optimal control laws for player P are:

$$\overset{*}{\varphi}(r,\theta) = \operatorname{sign} J_\theta(r,\theta)\ , \qquad \text{for } J_\theta(r,\theta) \neq 0,$$
$$\overset{*}{\varphi}(r,\theta) = \text{ any admissible } \varphi\ , \quad \text{for } J_\theta(r,\theta) = 0, \tag{7.51}$$

and for player E:

$$(\sin \overset{*}{\psi}(r,\theta), \cos \overset{*}{\psi}(r,\theta)) \parallel (J_r(r,\theta), \frac{1}{r} J_\theta(r,\theta)). \tag{7.52}$$

Let us substitute the optimal controls in ME1 to obtain ME2,

$$-\,|J_\theta(r,\theta)| + \beta\sqrt{J_r^2(r,\theta) + \frac{1}{r^2} J_\theta^2(r,\theta)} = 0. \tag{7.53}$$

ME1 is *homogeneous* in J_r and J_θ since $L(r,\theta,\varphi,\psi) = 0$. Hence, by equation 7.52, we can substitute $\cos \overset{*}{\psi}(r,\theta)$ for $J_r(r,\theta)$ and $\sin \overset{*}{\psi}(r,\theta)$ for $J_\theta(r,\theta)$ in ME2. We get:

$$\beta - \overset{*}{\varphi}\, r \sin \overset{*}{\psi} = 0. \tag{7.54}$$

It follows that:

$$\sin \overset{*}{\psi} = \overset{*}{\varphi} \frac{\beta}{r}. \tag{7.55}$$

There is a clear geometric interpretation to relation (7.55). *Lady* should swim along the tangent to a circle with radius β centered at the center of the pond! Equation (7.55) also indicates that we have candidate optimal control laws only for points (r,θ) where $r > \beta$! This is a sign that the *regular procedure* may fail to cover the game-set with candidate optimal trajectories!

The adjoint equations

$$\overset{\circ}{J}_r(r,\theta) = \tfrac{1}{r^2} J_\theta(r,\theta)\beta \sin \overset{*}{\psi},$$
$$\overset{\circ}{J}_\theta(r,\theta) = 0. \tag{7.56}$$

The tangent to the target-set at $(1,\theta)$ is

$$\mathbf{t} = \hat{\mathbf{e}}_\theta,$$

where ($\hat{\mathbf{e}}_\theta$ is a unit vector in the θ-direction).
Notice that:

$$\nabla G(1,\theta) = \mathbf{e}_\theta\ , \qquad \text{for } 0 < \theta < \pi,$$
$$\nabla G(1,\theta) = -\mathbf{e}_\theta\ , \qquad \text{for } -\pi \leq \theta < 0. \tag{7.57}$$

Using the requirement

$$\nabla J(1,\theta)\cdot \mathbf{t} = \nabla G(1,\theta)\cdot \mathbf{t},$$

we obtain:

$$\begin{aligned} J_\theta(1,\theta) &= 1\ , \qquad \text{for } 0<\theta<\pi, \\ J_\theta(1,\theta) &= -1\ , \qquad \text{for } -\pi \le \theta < 0. \end{aligned} \tag{7.58}$$

We can specialize ME2 (equation (7.53)) to the point $(1,\theta)$,

$$1+\beta\sqrt{J_r^2(1,\theta)+\frac{1}{r^2}}=0, \tag{7.59}$$

and get an expression for $J_r(1,\theta)$:

$$J_r(1,\theta)=\sqrt{\frac{1}{\beta^2}-1}. \tag{7.60}$$

From equations (7.56) and equation (7.53) we can get by substituting $|J_\theta(r,\theta)|=1$ an explicit expression for $J_r(r,\theta)$[13]:

$$J_r(r,\theta)=\sqrt{\frac{1}{\beta^2}-\frac{1}{r^2}}. \tag{7.61}$$

Once more we realize that our construction is going to breakdown for $r<\beta$.

The optimal trajectories

We find that the interior of the circle $r=\beta$ is not covered by candidate optimal trajectories and that some tarjectories from points on the target-set having positive θ intersect with trajectories that start with negative θ.

Again we have to assume the existence of a singular surface in order to proceed and complete the solution.

In the case of this game we can resolve those difficulties using our intuition. We observe that as long as $r<\beta$ *Lady* dominates $\dot\theta$. This means that she can get as close as he wishes to the point $(\beta,0)$ and guarantee herself *at least* the outcome of a play that starts there[14].

The *value map* (the map of the iso-value lines) of the game is given in Figure 7.3. We see that most of the game-set is a *Plateu Surface* meaning that the same value (corresponding to the value at the point $(\beta,0)$) is guaranteed. We also have (even from considerations of symmetry) that along the symmetry line $\theta=0$ two fields of optimal trajectories intersect[15]. The line

---

[13] This is an example for the use of ME2 as a *first integral* as mentioned in section 7.2.3.

[14] In the sense of the extended solution concept discussed in section 3.3.

[15] We also notice that ∇J is *discontinuous* across this line.

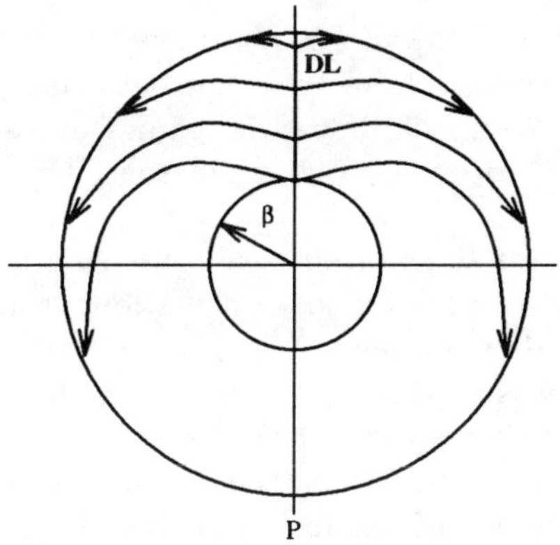

FIGURE 7.3. Value map for the Lady in the Lake problem

$\theta = 0$ is a locus of points where player P can choose whether the optimal trajectory will proceed right or left. We shall discuss this kind of line in chapter 11.

Remarks
The solution of the *Lady in the Lake* game demonstrates the following important phenomena:

1. There can be a region of the *UP* where optimal trajectories do not terminate. So in general we cannot assume ahead of the completion of a solution that at every point of the *UP* $G(\mathbf{x}) = J(\mathbf{x})$!

2. A part of the regular field of candidate optimal trajectories constructed backward from the *UP* was discarded as the singular phenomenon was introduced.

•

7.4 Linear Quadratic Games

7.4.1 Introduction

The name *linear quadratic differential games*[16] refers to the fact that in this family of games the *equation of motion* is a *linear* function of the state $\mathbf{x}$ and of the controls $\mathbf{u}$ and $\mathbf{v}$ and that the *outcome functional* is a function of *quadratic forms* of $\mathbf{x}$, $\mathbf{u}$ and $\mathbf{v}$.

[16] We shall often use *LQG* as an abbreviation for *linear quadratic differential games*.

People were attracted to the use of the linear quadratic formulation for modelling real life problems[17] (in the context of control theory as well as in the context of the theory of differential games) in the early stages of the development of the theories for various reasons. One reason may be the availability of many theoretical results from the theory of *linear algebra* and the theory of *linear differential equations*. Another reason is because quadratic cost functions *penalize* large deviations from a desired *nominal* much more than small deviations.
A LQG formulation was used by *Y. C. Ho, A. E. Bryson* and *S. Baron* in [ABB65] where it was shown that for games of finite duration the existence of a solution to a Riccati equation (the absence of a *conjugate point*) is sufficient for the existence of a solution of the game. In [Ber79] *P. Bernhard* studied the issue of the existence of a solution for LQG with a wider class of *target-sets*. He also showed that it is possible that a LQG will have a solution even if there exists a conjugate point. *E. F. Mageirou* investigated LQG with *infinite horizon* in [Mag76]. The idea that LQG can be utilized in the analysis and design of *controllers* was mentioned by some researchers (see *E. F. Mageirou* and *Y. C. Ho* in [MH77]). In the recent years the theory of *linear quadratic differential games* (with finite duration and with infinite horizon) was linked with the analysis and worst case design of controllers. The reader is referred to the works of *J. Doyle, K. Glover, P. Khargonekar* and *B. Francis* in [JDF89], *T. Basar* in [Bas90], *T. Basar* and *P. Bernhard* in [BB91] and *J. L. Speyer* in [SR91].

In this section we shall first derive the basic results for the case of a finite duration LQG with unbounded controls. Next we investigate what happens when the duration of the game increases indefinitely. The last part deals with the basic correlation of LQG to optimal controllers. The purpose of this section is to acquaint the reader with the basics of the theory. The more advanced applications of the results of the theory of LQG to controller design problems are outside the scope of this book.

7.4.2 LQG with Fixed Duration and Unbounded Controls

Consider the following *linear quadratic differential game*:

The game-set:

$$\mathcal{S} \equiv \{(x,t) \mid \; x \in R^n \;,\; t_0 \le t \le T\}.$$

[17] A LQG model may be used for the analysis of the effects of small perturbations from a nominal in nonlinear problems.

The target-set:

$$\mathcal{C} \equiv \{(x,t) \mid x \in R^n \ , \ t = T\}.$$

The equations of motion:

$$\dot{x} = Fx + Gu + Ev \ , \quad x(t_0) = x_0,$$

$$\dot{t} = 1.$$

Notice that we consider the time t as a state variable.

The outcome functional:

$$\pi_T[x,t,u(\cdot),v(\cdot)] = x^T(T)Q(T)x(T) + \int_t^T (x^T Qx + u^T Ru - v^T Bv)d\tau.$$

Q, R, B are symmetric matrices.

The *Hamiltonian function* is:

$$H(x,t,J_x,J_t,u,v) = J_x(Fx + Gu + Ev) + J_t + (x^T Qx + u^T Ru - v^T Bv),$$

and the *Isaacs equations* have the following form:

$$J = J(x,t),$$

$$\underset{\forall v \in \mathcal{V}}{H(x,t,J_x,J_t,\overset{*}{u},v)} \leq H(x,t,J_x,J_t,\overset{*}{u},\overset{*}{v}) = 0 \leq \underset{\forall u \in \mathcal{U}}{H(x,t,J_x,J_t,u,\overset{*}{v})}. \tag{7.62}$$

Candidate *optimal control laws* that satisfy *Isaacs Equations* can be derived using the condition:

$$\frac{\partial H(x,t,J_x,J_t,u,v)}{\partial u} = 0,$$

$$\frac{\partial H(x,t,J_x,J_t,u,v)}{\partial v} = 0.$$

We get:

$$\overset{*}{u} = -\tfrac{1}{2}R^{-1}G^T {J_x}^T,$$

$$\overset{*}{v} = +\tfrac{1}{2}B^{-1}E^T {J_x}^T.$$

Let us assume that the *value function* of the game has the form

$$J(x,t) = x^T S(t,T)x. \tag{7.63}$$

In this case the *optimal control* can be stated in a *closed loop* form:

$$\begin{aligned} \overset{*}{u} &= -R^{-1}G^T Sx, \\ \overset{*}{v} &= +B^{-1}E^T Sx. \end{aligned} \tag{7.64}$$

In order to satisfy ME2 it is necessary that $S(t,T)$ should satisfy the following *Riccati matrix differential equation*[18]

$$\begin{array}{l}\dot{S} + SF + F^T S - S(GR^{-1}G^T - EB^{-1}E^T)S + Q = 0. \\ S(T,T) = Q(T)\end{array} \tag{7.65}$$

Lemma 7.4.1 (Completing the squares)

Suppose that $S(t,T)$ is the solution of the *Riccati* equation of a finite time linear quadratic differential game played over the time interval $[t_0, T]$. then for any pair of control-histories[19] $(u(\cdot), v(\cdot))$,

$$\begin{array}{l}\pi_T[x,t,u(\cdot),v(\cdot)] = x^T(t)S(t,T)x(t) + \\ + \int_t^T \{\|u + R^{-1}G^T S(\tau,T)x\|_R^2 - \|v - B^{-1}E^T S(\tau,T)x\|_B^2\} d\tau.\end{array} \tag{7.66}$$

Proof

Integrating the time rate of $J(x,t)$ along an *optimal path* and using equations (7.63) and (7.64) we get:

$$\begin{array}{l}x^T(T)S(T,T)x(T) - x^T(t)S(t,T)x(t) = \int_t^T \frac{d}{d\tau}[J(x,\tau)]d\tau = \\ = \int_t^T \{2x^T S[Fx + Gu + Ev] + x^T \dot{S} x\} d\tau.\end{array} \tag{7.67}$$

Integrating the *Riccati equation* (7.65) we get:

$$0 = -\int_t^T x^T[\dot{S} + SF + F^T S - S(EB^{-1}E^T - GR^{-1}G^T)S + Q]x \, d\tau. \tag{7.68}$$

From the definition of the *outcome functional* we have that:

$$\pi_T[x,t,u(\cdot),v(\cdot)] = x^T(T)Q(T)X(T) + \int_t^T \{x^T Q x + u^T R u - v^T B v\} d\tau. \tag{7.69}$$

Adding equations (7.67), (7.68) and (7.69) and recalling that

$$S(T,T) = Q(T),$$

---

[18] In the derivation we made use of the property of *quadratic forms* to substitute

$$2x^T SFx = x^T(SF + F^T S)x,$$

[19] The definition of the norms are:

$$\|u + R^{-1}G^T S(\tau,T)x\|_R^2 = [u + R^{-1}G^T S(\tau,T)x]^T R[u + R^{-1}G^T S(\tau,T)x],$$

$$\|v - B^{-1}E^T S(\tau,T)x\|_B^2 = [v - B^{-1}E^T S(\tau,T)x]^T B[v - B^{-1}E^T S(\tau,T)x].$$

the lemma follows.

•

If a solution $S(t,T)$ of the *Riccati equation* (7.65) exists for $t_0 \leq t \leq T$ we can use equation (7.66) for a direct proof that the *solution triplet*

$$[J(x,t) = x^T S(t,T)x,\ \overset{*}{u} = -R^{-1}G^T Sx,\ \overset{*}{v} = B^{-1}E^T Sx],$$

satisfies the *saddle* relation:

$$\pi_T[x,t,\overset{*}{u}(\cdot),v(\cdot)] \leq \ \pi_T[x,t,\overset{*}{u}(\cdot),\overset{*}{v}(\cdot)] = J(x,t) \leq \ \pi_T[x,t,u(\cdot),\overset{*}{v}(\cdot)].$$

Notice that those results remain valid also for the cases where the matrices F,G,E,Q,R,B are time-varying.

Example 7.4.1

Let us consider the following simple example:[20]

$$\dot{x} = x + u + v,$$

$$\dot{t} = 1,$$

$$\pi_T[x,t,u(\cdot),v(\cdot)] = \int_t^T (x^2 + u^2 - 2v^2)d\tau.$$

Assume the *value function* to be of the form:

$$J(x,t) = x^T S(t,T)x.$$

The Riccati equation for this game is:

$$\dot{S} = \tfrac{1}{2}S^2 - 2S - 1, \quad S(T,T) = 0.$$

The solution can be given in a closed form as

$$S(t,T) = \sqrt{6}\tanh[\sqrt{6}(t-T) + \tanh^{-1}(-2/\sqrt{6})] + 2.$$

•

[20] This example is due to *E. F. Mageirou*, see [Mag76].

7.4.3 Infinite Horizon Linear Quadratic Games

When we speak of *infinite horizon* we mean that we let $T \to \infty$. In order to investigate the meaning of *linear quadratic games with infinite horizon* we first consider *finite time linear quadratic differential game* played over the time interval $[t_0, T]$ with no terminal cost term in the outcome functional and next we let the final time $T \to \infty$.

The game set:

$$\mathcal{S} \equiv \{(x,t) \mid x \in R^n \ , \ t_0 \le t \le T\}.$$

The target set:

$$\mathcal{C} \equiv \{(x,t) \mid x \in R^n \ , \ t = T\}.$$

The equations of motion:

$$\dot{x} = Fx + Gu + Ev \qquad x(t_0) = x_0,$$

$$\dot{t} = 1.$$

The Outcome Functional:

$$\pi_T[x,t,u(\cdot),v(\cdot)] = \int_t^T (x^T Qx + u^T Ru - v^T Bv)d\tau.$$

Q, R, B are symmetric.

Let us first establish some results concerning *estimations* of the *value function* of this game.

Lemma 7.4.2

Let $t > \Delta > 0$. If $S(t,T)$ is non negative definite in the interval $(t_0 - \Delta, T)$,

$$x^T S(t-\Delta, T)x \ge x^T S(t,T)x \quad \text{for all} \quad t_0 \le t \le T,$$

and for $\Delta > 0$ if $S(t,T)$ is non negative definite in the interval $(t_0, T+\Delta)$,

$$x^T S(t, T+\Delta)x \ge x^T S(t,T)x \quad \text{for all} \quad t_0 \le t \le T.$$

Proof

The function

$$V(x,t) = x^T S(t-\Delta, T)x,$$

and the *strategies*

$$\tilde{u} = -R^{-1}G^T S(t-\Delta, T)x,$$

$$\tilde{v} = +B^{-1}E^T S(t-\Delta, T)x,$$

constitute an *estimating function* (see Lemma 6.4.1) because:

1. The ME1 and ME2 equations in the $[t-\Delta, T]$ game assures:

$$H(x,t,V_x,V_t,\tilde{u},v) \leq \; H(x,t,V_x,V_t,\tilde{u},\tilde{v}) = 0, \quad \forall v \in \mathcal{V}$$

$$0 = H(x,t,V_x,V_t,\tilde{u},\tilde{v}) \leq \; H(x,t,V_x,V_t,u,\tilde{v}). \quad \forall u \in \mathcal{U}$$

2. Termination in finite time is assured (invariantly!).

3. Since $S(t-\Delta, T)$ is assumed to be non-negative-definite

$$V(x,T) = x^T S(T-\Delta,T)x \geq 0 = J(x,T).$$

Hence for all $x, t \geq \Delta$

$$x^T S(t,T)x = J(x,t) \leq V(x,T) = x^T S(t-\Delta,T)x.$$

Similarly, The function

$$V(x,t) = x^T S(t, T+\Delta)x,$$

and the *strategies*

$$\tilde{u} = -R^{-1}G^T S(t,T+\Delta)x,$$

$$\tilde{v} = +B^{-1}E^T S(t,T+\Delta)x,$$

constitute an *estimating function* (see Lemma 6.4.1) because:

1. The ME1 and ME2 equations in the $[t, T+\Delta]$ game assures:

$$H(x,t,V_x,V_t,\tilde{u},v) \leq \; H(x,t,V_x,V_t,\tilde{u},\tilde{v}) = 0, \quad \forall v \in \mathcal{V}$$

$$0 = H(x,t,V_x,V_t,\tilde{u},\tilde{v}) \leq \; H(x,t,V_x,V_t,u,\tilde{v}). \quad \forall u \in \mathcal{U}$$

2. Termination in finite time is assured (invariantly!).

3. Since $S(t, T+\Delta)$ is assumed to be non-negative-definite

$$V(x,T) = x^T S(T,T+\Delta)x \geq 0 = J(x,T).$$

Hence for all $x, t \geq T$

$$x^T S(t,T)x = J(x,t) \leq V(x,t) = x^T S(t, T+\Delta)x.$$

•

Let us now restrict ourselves to the class of games where the matrices F, G, E, Q, R, B are also time invariant.
We now have:

$$S(t,T) = S(T-t).$$

Lemma 7.4.3

If S^* is a non negative-definite solution of [21]

$$SF + F^T S + Q - S(GR^{-1}G^T - EB^{-1}E^T)S = 0, \tag{7.70}$$

we have that for all (t, T)

$$x^T(t)S^* x(t) \geq x^T(t)S(t,T)x(t).$$

Proof
The function

$$V(x,t) = x^T S^* x,$$

and the *strategies*

$$\tilde{u} = -R^{-1}G^T S^* x,$$

$$\tilde{v} = +B^{-1}E^T S^* x.$$

constitute an *estimating function* (see Lemma 6.4.1) because:

1. The choice of controls and the *algebraic Riccati equation* assures:

$$H(x,t,V_x,V_t,\tilde{u},v) \leq \underset{\forall v \in \mathcal{V}}{} H(x,t,V_x,V_t,\tilde{u},\tilde{v}) = 0,$$

$$0 = H(x,t,V_x,V_t,\tilde{u},\tilde{v}) \leq \underset{\forall u \in \mathcal{U}}{H(x,t,V_x,V_t,u,\tilde{v})}.$$

2. Termination in finite time is assured (invariantly!).

---

[21] Equation (7.70) is called the *algebraic Riccati equation (ARE)* because it can be obtained from the Riccati equation (7.65) by setting $\dot{S} = 0$.

3. Since S^* is assumed to be non-negative-definite

$$V(x,T) = x^T S^* x \geq 0 = J(x,T).$$

Hence for all $t_0 \leq t \leq T$,

$$x^T S(t,T)x = J(x,t) \leq V(x,t) = x^T S^* x.$$

•

The properties of $[V(x,t), \tilde{u}, \tilde{v}]$ enable us to *complete the squares*:

Corollary 7.4.4

If S^* is a non negative definite solution of the *algebraic Riccati equation* for a finite time linear quadratic differential game played over the time interval $[t_0, T]$, then for any pair of control-histories $(u(\cdot), v(\cdot))$

$$\begin{aligned}\pi[x,t,T,u(\cdot),v(\cdot)] &= x^T(t)S^*x(t) - x^T(T)S^*x(T) + \\ &+ \textstyle\int_t^T \{\|u + R^{-1}G^T S^* x\|_R^2 - \|v - B^{-1}E^T S^* x\|_B^2\} d\tau.\end{aligned} \tag{7.71}$$

Proof

Integrating the complete differential we get:

$$\begin{aligned}x^T(T)S^*x(T) - x^T(t)S^*x(t) &= \textstyle\int_t^T \frac{d}{d\tau}[V(x,\tau)]d\tau = \\ &= \textstyle\int_t^T \{2x^T S^*[Fx + Gu + Ev]\} d\tau.\end{aligned} \tag{7.72}$$

Integrating the *algebraic Riccati equation* we get:

$$0 = -\int_t^T x^T[SF + F^T S - S(EB^{-1}E^T - GR^{-1}G^T)S + Q]x \, d\tau. \tag{7.73}$$

Recall the definition:

$$\pi_T[x,t,u(\cdot),v(\cdot)] = \int_t^T \{x^T Qx + u^T Ru - v^T Bv\} d\tau. \tag{7.74}$$

Adding equations (7.72), (7.73) and (7.74) the lemma follows.

•

Let us consider an infinite sequence of finite time linear quadratic games with no terminal cost played over the time interval (t_0, T) where $T \to \infty$ (see [Mag76]). We want to investigate the existence and the meanings of the limit:

$$\lim_{T \to \infty} x^T S(T-t)x = J^\infty(x,t),$$

which is equivalent to:[22]

$$\lim_{T\to\infty} S(T-t) = S^+.$$

In the following we shall denote

$$\tau = T - t.$$

Lemma 7.4.5

$\lim_{\tau\to\infty} S(\tau) = S^+$ exists.

Proof
Lemma 7.4.2 provides for $\Delta > 0$:

$$x^T S(\tau+\Delta)x \geq x^T S(\tau)x,$$

and Lemma 7.4.3 gives that for all τ

$$x^T S^* x \geq x^T S(\tau)x.$$

As $\tau \to \infty$ the positive quantity $x^T S(\tau)x$ increases monotonously but remains bounded from above by $x^T S^* x$ so it must converge to a unique limit $x^T S^+ x$. S^+ is hence non negative-definite matrix.

•

Lemma 7.4.6

The matrix S^+ is the smallest non negative-definite solution of the *algebraic Riccati equation*.

Proof

$$\frac{dS(\tau)}{d\tau} = -\frac{\partial S(t,T)}{\partial t},$$

so $S(\tau)$ satisfies the *Riccati equation*

$$-\dot{S} + SF + F^T S + Q - S(GR^{-1}G^T - EB^{-1}E^T)S = 0. \tag{7.75}$$

Obviously,

$$\lim_{\tau\to\infty} \frac{dS(\tau)}{d\tau} = 0,$$

---

[22] For the case where F, G, E, Q, R, B are time invariant we obviously have

$$S(t,T) = S(T-t).$$

while all the other quantities in equation (7.75) remain bounded so $S(\tau)$ converges to a solution of

$$SF + F^T S + Q - S(GR^{-1}G^T - EB^{-1}E^T) = 0. \qquad (7.76)$$

If $S^\dagger < S^+$ is another non negative-definite solution of the *algebraic Riccati equation* (7.76) then by lemma 7.4.3 we have that for all τ

$$x^T S^\dagger x \geq x^T S(\tau)x,$$

which implies,

$$x^T S^\dagger x \geq \lim_{\tau\to\infty} x^T S(\tau)x = x^T S^+ x,$$

contradicting the assumption on $S^\dagger$.

•

Lemma 7.4.7

If the matrix Q is positive definite then also S^+ is positive definite [23].

Proof

We have already shown that S^+ need to be non negative-definite. To prove the lemma it will suffice to show that that any symmetric matrix solution of the *algebraic Riccati equation* is invertible.

Suppose that $S^\ddagger$ is a symmetric noninvertible solution of equation (7.76). Then for some $x \neq 0$ we shall have $S^\ddagger x = 0$.
Let us substitute $S^\ddagger$ in equation (7.76) and premultiply and postmultiply it by x^T and x. We shall get that for some $x \neq 0$ we have $x^T Qx = 0$ contradicting the assumption that the matrix Q is positive definite.

•

Let us follow *E. F. Mageirou*[Mag76] and use Example 7.4.1 to show the remarkable fact that for the limit strategies:

$$\hat{u} = -R^{-1}G^T S^+ x, \qquad \hat{v} = +B^{-1}E^T S^+ x, \qquad (7.77)$$

The candidate triplet $[J^\infty(x), \hat{u}, \hat{v}]$ fails to satisfy the saddle relation

$$\lim_{T\to\infty} \pi_T[x,t,\hat{u}(\cdot),v(\cdot)] \leq \lim_{T\to\infty} \pi_T[x,t,\hat{u}(\cdot),\hat{v}(\cdot)] = J^\infty(x), \quad \forall v(\cdot)$$

$$\lim_{T\to\infty} \pi_T[x,t,\hat{u}(\cdot),\hat{v}(\cdot)] = J^\infty(x) \leq \lim_{T\to\infty} \pi_T[x,t,u(\cdot),\hat{v}(\cdot)]. \quad \forall u(\cdot) \qquad (7.78)$$

[23] The proof is due to *E. F. Mageirou*, see [Mag76].

Example 7.4.2

Let us continue with Example 7.4.1 and check what happens as $T \to \infty$. Clearly:

$$lim_{T\to\infty} S(t,T) = 2+\sqrt{6},$$

and

$$\hat{u} = -(2+\sqrt{6})x,$$

$$\hat{v} = \tfrac{1}{2}(2+\sqrt{6})x,$$

$$\pi_T[x,t,u(\cdot),\hat{v}(\cdot)] = \int_t^T \{x^2+u^2-2[\frac{1}{2}(2+\sqrt{6})x]^2\}dt.$$

It is easy to see that player P benefits if he deviates from $\hat{u}$ and utilizes $u=0$. (The outcome will decrease indefinitely).

•

Lemma 7.4.8

Player P and player E have ϵ-strategies u_ϵ, v_ϵ that can guarantee

$$\pi_T[x,t,u_\epsilon(\cdot),v(\cdot)] \leq x^T S^+ x + \epsilon,$$

$$\pi_T[x,t,u(\cdot),v_\epsilon(\cdot)] \geq x^T S^+ x - \epsilon,$$

Proof
Recall Lemma 7.71 and substitute $u = u_\epsilon = -R^{-1}G^T S^+ x$. We obtain that for all T:

$$\begin{aligned}\pi_T[x,t,u_\epsilon(\cdot),v(\cdot)] &= x^T S^+ x - x^T(T)S^+ x(T)+ \\ &+ \textstyle\int_t^T (-\|v - B^{-1}E^T S^+ x\|_B^2)d\tau,\end{aligned} \tag{7.79}$$

or equivalently:

$$\begin{aligned}\pi_T[x,t,u_\epsilon(\cdot),v(\cdot)] &= J^\infty(x) - x^T(T)S^+ x(T)+ \\ &+ \textstyle\int_t^T (-\|v - B^{-1}E^T S^+ x\|_B^2)d\tau \leq J^\infty(x).\end{aligned} \tag{7.80}$$

Hence

$$\lim_{T\to\infty} \max_{v(\cdot)} \pi_T[x,t,u_\epsilon(\cdot),v(\cdot)] =\leq J^\infty(x). \tag{7.81}$$

Now what about player E?
Recall that $S(t,T)$ decreases monotonously with t and that

$$\lim_{T\to\infty} x^T S(t_0,T)x = J^\infty(x).$$

So given any $\epsilon > 0$ there exists $T(\epsilon)$ such that

$$J^{\infty}(x) - x^T S(t,T)x \leq \epsilon \quad \text{for all } t_0 \leq t \leq T(\epsilon),$$

player E may use the following ϵ-strategies:

$$v_\epsilon = -B^{-1}E^T S[t,T(\epsilon)]x \qquad \text{for all } t \leq T(\epsilon),$$

$$v_\epsilon = 0 \qquad \text{for all } t > T(\epsilon).$$

For all $T > T(\epsilon)$, we can split the outcome:

$$\pi_T[x,t,u(\cdot),v(\cdot)] = \pi_{T(\epsilon)}[x,t,u(\cdot),v(\cdot)] + \pi_T[x,T(\epsilon),u(\cdot),v(\cdot)].$$

Now by applying the lemma about *completing the squares*, (Lemma 7.4.1) we obtain:

$$\pi_T[x,t,u(\cdot),v(\cdot)] = x^T(t)S(t,T)x(t)+ \\ + \int_t^T \{\|u + R^{-1}G^T S(\tau,T)x\|_R^2 - \|v - B^{-1}E^T S(\tau,T)x\|_B^2\}d\tau. \tag{7.82}$$

So for all $T > T(\epsilon)$,

$$\pi_T[x,t,u(\cdot),v_\epsilon(\cdot)] = x^T(t)S(t,T)x(t)+ \\ + \int_t^{T(\epsilon)} \|u + R^{-1}G^T S(\tau,T)x\|_R^2 d\tau + \int_{T(\epsilon)}^T \{x^T Q x + u^T R u\}d\tau \geq \\ \geq x^T S(t,T)x \geq J^{\infty}(x) - \epsilon. \tag{7.83}$$

Obviously,

$$\lim_{T\to\infty} \min_{u(\cdot)} \pi_T[x,t,u(\cdot),v_\epsilon(\cdot)] \geq J^{\infty}(x) - \epsilon. \tag{7.84}$$

Thus completing the proof of the lemma.

•

Corollary 7.4.9

If the *algebraic Riccati equation* has nonnegative definite solutions then for ϵ-optimal trajectories

$$0 \leq x^T(T)S^+x(T) \leq \epsilon\ , \quad \text{for all } \ T > T(\epsilon)$$

Proof

The proof can be done by the arguments of the proof of Lemma 7.4.8 and is left to the reader.

•

7.4.4 LQG and Controller Design

Let us consider the following family[24] of *linear quadratic games* with respect to the two parameters γ and T:

The Equations of Motion:

$$\begin{aligned} \dot{x} &= Fx + Gu + Ev \ , \quad x(t_0) = x_0, \\ \dot{t} &= 1. \end{aligned} \tag{7.85}$$

The Outcome Functional:

$$\pi_{\gamma T}[x, t, u(\cdot), v(\cdot)] = x^T(T)Q(T)x(T) + \int_t^T (x^T Q x + u^T u - \gamma^2 v^T v) d\tau.$$

Q is a symmetric positive definite matrix. The system is assumed to be time varying[25].

Let Γ_T be the set of all $\gamma > 0$ such that in the whole interval $[0, T]$ the *Riccati equation*,

$$\begin{aligned} &\dot{S} + SF + F^T S - S(GG^T - \tfrac{1}{\gamma^2} EE^T)S + Q = 0, \\ &S(T) = Q(T) \end{aligned} \tag{7.86}$$

does not have a *conjugate point* so that a solution $S_\gamma(t, T)$ exists.
The *solution triplet* for a game with the particular parameters $(\gamma \in \Gamma, T)$ is:

$$[J_{\gamma T}, \overset{*}{u}_{\gamma T}, \overset{*}{v}_{\gamma T}].$$

The optimal controls are:

$$\begin{aligned} \overset{*}{u}_{\gamma T}\ (x, t) &= -G^T(t) S_\gamma(t, T) x, \\ \overset{*}{v}_{\gamma T}\ (x, t) &= \tfrac{1}{\gamma^2} E^T(t) S_\gamma(t, T) x. \end{aligned} \tag{7.87}$$

The *value function* is given by

$$J_{\gamma T}(x, t) = x^T S_\gamma(t, T) x.$$

Let us assume $t = 0$. We denote:

$$\pi_{x(0)T}[u(\cdot), v(\cdot)] = x^T(T)Q(T)x(T) + \int_0^T (x^T Q x + u^T u) d\tau \geq 0,$$

---

[24] This family of games is based on the LQG of section 7.4.2 with:

$$R = I \ , \quad B = \gamma^2 I.$$

[25] We shall often omit the time dependence in order to simplify the notations.

so

$$\pi_{\gamma T}[x, 0, u(\cdot), v(\cdot)] = \pi_{x(0)T}[u(\cdot), v(\cdot)] - \gamma^2 \int_0^T v^T v dt.$$

The *saddle property* of the solution of the differential game that starts at $x(0)$ gives:

$$\begin{aligned} \pi_{x(0)T}[\overset{*}{u}_{\gamma T}(\cdot), v(\cdot)] - & \gamma^2 \textstyle\int_0^T v^T v dt \leq \\ & \leq \pi_{x(0)T}[\overset{*}{u}_{\gamma T}(\cdot), \overset{*}{v}_{\gamma T}(\cdot)] - \gamma^2 \textstyle\int_0^T \overset{*}{v}_{\gamma T} \overset{*}{v}_{\gamma T} dt = \\ = x^T(0) S_\gamma(0,T) x(0) & \leq \pi_{x(0)T}[u(\cdot), \overset{*}{v}_{\gamma T}(\cdot)] - \gamma^2 \textstyle\int_0^T \overset{*}{v}_{\gamma T} \overset{*}{v}_{\gamma T} dt. \end{aligned} \tag{7.88}$$

•

Relation (7.88) indicates that the feedback control law $\overset{*}{u}_{\gamma T}$ of the *linear quadratic game* also solves the following *controller design problem*[26] over the time interval $[0, T]$:

The system equations:

$$\begin{aligned} \dot{x} &= Fx + Gu + Ev, \quad x(t_0) = x_0, \\ \dot{t} &= 1. \end{aligned} \tag{7.89}$$

The disturbance: The disturbance $v(\cdot)$ is any square integrable function, i.e. $v(\cdot) \in \mathcal{H}_v$. The L^2 norm of the *disturbance* is $\int_0^T v^T v dt$.

The controlled output: $\pi_{x(0)}^T[\overset{*}{u}_{\gamma T}(\cdot), v(\cdot)]$ is the L^2 norm of the *controlled outout*.

The objective: The objective is to synthesize a control law $\tilde{u}_{\gamma T}$ to guarantee that the disturbances are *attenuated* in a way that satisfies

$$\pi_{x(0)T}[\overset{*}{u}_{\gamma T}(\cdot), v(\cdot)] \leq \gamma^2 \int_0^T v^T v dt + x^T(0) S_\gamma(0,T) x(0). \quad \text{for all } v(\cdot) \in \mathcal{H}_v \tag{7.90}$$

We argue (heuristically) that for an arbitrarily small $\epsilon > 0$ relation (7.88) provides that for $x(0) \to 0$

$$\pi_{0T}[\overset{*}{u}_{\gamma T}(\cdot), v(\cdot)] \leq (1+\epsilon)\gamma^2 \int_0^T v^T v dt, \quad \text{for any } v(\cdot) \in \mathcal{H}_v, \tag{7.91}$$

so:

$$\sup_{v(\cdot) \in \mathcal{H}_v} \pi_{0T}[\overset{*}{u}_{\gamma T}(\cdot), v(\cdot)] = \gamma^2 \int_0^T v^T v dt. \tag{7.92}$$

---

[26]The formulation resembles the controllers discussed by *K. Uchida* and *M. Fujita* in [UF90].

•

Relation (7.92) indicates that the feedback control law $\overset{*}{u}_{\gamma T}$ of the *linear quadratic game* also solves a *controller design problem* over the time interval $[0, T]$:
The controlled system is as before, the initial state is $x(0) = 0$ and the objective is to synthesize a control law $\tilde{u}_{\gamma T}$ to guarantee that the disturbances are attenuated in a way that satisfies

$$\begin{array}{l} \pi_{0T}[\overset{*}{u}_{\gamma T}(\cdot), v(\cdot)] \leq \gamma^2 \int_0^T v^T v dt. \\ \text{for all } v(\cdot) \in \mathcal{H}_v \end{array} \tag{7.93}$$

•

If we define[27]

$$M_T[u(\cdot), v(\cdot)] = \frac{\{\pi_{0T}[u(\cdot), v(\cdot)]\}^{\frac{1}{2}}}{[\int_0^T v^T v dt]^{\frac{1}{2}}}. \tag{7.94}$$

We can use the definition (7.94) to rewrite relation (7.92) and to obtain a correlation between the LQG and a different formulation of the *controller design problem*[28]

$$\begin{array}{l} \sup M^T[\tilde{u}_{\gamma T}(\cdot), v(\cdot)] = \gamma. \\ v(\cdot) \in \mathcal{H}_v \end{array} \tag{7.95}$$

Relation (7.95) indicates that the optimal control law $\overset{*}{u}_{\gamma T}$ of the LQG with the parameters (γ, T) also solves the *γ-bounded disturbance-gain controller design-problem* over the time interval $[0, T]$ where the objective is to synthesize a control law $\tilde{u}_{\gamma T}$ to guarantee that the *gain* $M^T[\tilde{u}_{\gamma T}(\cdot), v(\cdot)]$ defined in equation (7.94) will not exceed γ for all $v(\cdot) \in \mathcal{H}_v$.

•

Let $\overset{*}{\gamma}_T$ be the *upper lower bound* of the values of γ that belong to Γ_T. From relation (7.95) it follows that:

$$\begin{array}{ll} \inf & \sup M_T[\tilde{u}_{\gamma T}(\cdot), v(\cdot)] = \overset{*}{\gamma}_T. \\ \gamma \in \Gamma_T & v(\cdot) \in \mathcal{H}_v \end{array} \tag{7.96}$$

This corresponds to the problem of designing a *controller with minimum disturbance gain*[29].

•

If we consider the time-invariant version of the previously discussed family of *linear quadratic games* with the parameters (γ, T) and we let $T \to \infty$

---

[27] Assuming $v(\cdot) \neq o$.

[28] This formulation resembles the controllers discussed by *T. Basar* in [Bas90].

[29] As noted in [Bas90] we can actually synthesize controls that guarantee a gain that will not exceed $\overset{*}{\gamma}_T + \epsilon$ for arbitrarily small $\epsilon > 0$.

we obtain a family of *linear quadratic games with infinite horizon*[30]. Let Γ_∞ be the set of all $\gamma > 0$ such that the *algebraic Riccati equation*

$$SF + F^T S - S(GG^T - \frac{1}{\gamma^2} EE^T)S + Q = 0, \tag{7.97}$$

has a positive definite solution.
From the results of section 7.4.3 we have that for $\gamma \in \Gamma_\infty$ if S_γ^+ is the lowest positive definite solution of (7.97) then the *value function* is given by

$$J_\gamma^\infty(x,t) = x^T S_\gamma^+ x.$$

And given any $\epsilon > 0$ we have that the ϵ-optimal control for player P is:

$$u_\gamma^\epsilon = -G^T S_\gamma^+ x.$$

Recall that given any $\epsilon > 0$ there exists $T(\epsilon)$ such that

$$J_\gamma^\infty(x) - x^T S_\gamma^T(t,T)x \le \epsilon \quad \text{for all } t_0 \le t \le T(\epsilon),$$

Player E may use the following ϵ-strategies:

$$v_\gamma^\epsilon = -\tfrac{1}{\gamma^2} E^T S_\gamma^T[t, T(\epsilon)]x \qquad \text{for all } t \le T(\epsilon),$$

$$v_\gamma^\epsilon = 0 \qquad \text{for all } t > T(\epsilon).$$

It is not difficult[31] to show that this family of LQG with infinite horizon is correlated to the corresponding *infinite-horizon* version of the disturbance attenuation controller problems discussed previously.
Let $\overset{*}{\gamma}_\infty$ be the *upper lower bound* of the values of γ that belong to Γ_∞.
It follows, for example, that:

$$\inf_{\gamma \in \Gamma_\infty} \quad \sup_{v(\cdot) \in \mathcal{H}_v} M^\infty[\tilde{u}_\gamma^\infty(\cdot), v(\cdot)] = \overset{*}{\gamma}_\infty. \tag{7.98}$$

This corresponds to the problem of designing an *infinite-horizon controller with minimum disturbance gain*[32].

•

---

[30] This family of games is based on the LQG discussed in section 7.4.3 with:

$$R = I\ , \quad B = \gamma^2 I.$$

[31] Proper attention should be given in the arguments to $\lim_{T\to\infty}$ of the functionals .

[32] We can actually synthesize controls that guarantee a gain that will not exceed $\overset{*}{\gamma}_\infty + \epsilon$ for any arbitrary small $\epsilon > 0$. See *T. Basar* [Bas90].

7.5 Problems

Problem 7.5.1 (The Wall Pursuit game)

This problem was first described by *R. Isaacs* in [Isa75] An evader E (an escaping criminal ...) is constrained to run along a straight wall. A pursuer (the policeman ...), that has some speed advantage, chases the evader. Capture occurs when P gets within a distance l from E.

The game-set:

$$\mathcal{S} \equiv \{(x, y) \mid y \geq o \ , \ \ x^2 + y^2 \geq l^2\}.$$

The target-set:

$$\{(x, y) \mid y \geq o \ , \ \ x^2 + y^2 = l^2\}.$$

The equations of motion:

$$\dot{x} = \psi - w \cos\varphi, \quad |\psi| \leq 0,$$

$$\dot{y} = -w \sin\varphi,$$

$w > 1$ is a constant. Player P chooses φ and player E chooses ψ.

The outcome functional:

$$G(x, y) = 0 \ , \quad L(x, y, \varphi, \psi) = 1. \tag{7.99}$$

1. Solve the game.
2. Show that the *isochrones* are circular arcs.
3. Show that the *value* is:

$$J(x, y) = \frac{(|x| - wl) + \sqrt{(|x|w - l)^2 + y^2(w^2 - 1)}}{w^2 - 1}.$$

•

Problem 7.5.2

This problem is a variant of Problem 7.5.1.

The game-set:

$$\mathcal{S} \equiv \{(x, y) \mid x^2 + y^2 \leq R^2\}.$$

The target-set:

$$\{(x,y) \mid x^2+y^2=R^2\}.$$

The equations of motion:

$$\dot{x}=u_1+v, \quad u_1^2+u_2^2 \le w^2, \quad |v| \le 1,$$

$$\dot{y}=u_2,$$

$w<1$ is a constant. Player P chooses (u_1,u_2) and player E chooses v.

The outcome functional:

$$G(x,y)=0 \quad L(x,y,u_1,u_2,v)=1. \tag{7.100}$$

1. Solve the game.
2. Show that the optimal trajectories in the first quadrant are straight lines that pass through the point $(R/w,0)$.

•

Problem 7.5.3 (The Lady in the Lake)

Consider Example 7.3.4.

1. Redefine the game in three dimensions using a fixed frame centered at the center of the pond.
2. Let (x,y) be the coordinates of player E and let α be the position of player P on the circumference of the circular pond. Show that the optimal trajectory of E is a straight line.

•

Problem 7.5.4

Consider the game:

The game-set:

$$\mathcal{S} \equiv \{(x,y,z) \mid z \ge 0\}.$$

The target-set:

$$\mathcal{C} \equiv \{(x,y,z) \mid z \ge 0\}.$$

The equations of motion:

$$\dot{x} = y,$$

$$\dot{y} = u + v \ , \quad |u| \leq U_m \ , \ |v| \leq V_m,$$

$$\dot{z} = 1.$$

The outcome functional:

$$G(x, y, z) = x^2/2 \ , \quad L(x, y, z) = 0.$$

1. Show that this game and Example 7.3.3 are different formulation of the same problem.
   Hint: use the substitutions: $\tau = t_f - z$ and $r = x + (t_f - t)y$.
   See [GL76] for the association of this problem to missile guidance.

•

Problem 7.5.5 (The Bang-Bang-Bang problem)

The *Bang-Bang-Bang Problem* was introduced by *R. Isaacs* [Isa69] and discussed by *M. D. Ciletti* [Cil70], *D. J. Wilson*[Wil72] and *J. Lewin*[Lew76].

The game-set:

$$\mathcal{S} \equiv \{(x, y) \mid y \geq -1\}.$$

The target-set:

$$\mathcal{C} \equiv \{(x, y) \mid y = -1\}.$$

The equations of motion:

$$\dot{x} = -\varphi \tfrac{1}{2}\sqrt{3}y + \sqrt{3}(2 - \tfrac{1}{2}y) + 2\sqrt{3}\cos\psi \ , \quad |\varphi| \leq 1,$$

$$\dot{y} = \varphi(3 + \tfrac{1}{2}y) + (\tfrac{1}{2}y - 1) + 2\sqrt{3}\sin\psi.$$

Player P chooses φ and player E chooses ψ.

The outcome functional: The payoff is terminal:

$$G(x, y) = -x \ , \quad L(x, y) = 0.$$

Player P wants to terminate and wishes to minimize the outcome while player E wishes to maximize the outcome.

1. Construct the family of candidate optimal trajectories emanating backwards from the UP and show that it does not cover the entire game-set.

2. Conjecture that the optimal trajectories are

$$x = c \qquad \text{for } y \geq 0,$$

$$y = -x/\sqrt{3} + c/\sqrt{3} \quad \text{for } y < 0.$$

and show that the corresponding value function is valid only in the sense of the extended solution concept.
Suggest the ϵ-optimal control law for player P.

•

Problem 7.5.6 (The Obstacle Tag problem)

The *Obstacle Tag game* was first mentioned in Isaacs [Isa75]. The problem was discussed by *J. R. Isbell* [Isb67], *D. Leshem* [Les85] and by *J. V. Breakwell* [Bre]. The game consists of two players P, the pursuer, and player E, the evader, in a planar simple motion, which allows both players to have unlimited rate of turn. pursuer, P, is the faster player, chasing player E, The speeds of player P and player E are bounded by w and 1 respectively, with $w > 1$.
The pursuit takes place in a plane that contains a fixed circular obstacle into which neither player is allowed to enter.

We formulate the problem in the form of a 3-dimensional game (see Figure 2.3).

The game-set:

$$\mathcal{S} \equiv \{(r_p, r_e, \theta) \mid r_p \geq 1, r_e \geq 1\}.$$

The terget-set:

$$\mathcal{C} \equiv \{(r_p, r_e, \theta) \mid r_p - r_e = 0\}.$$

The equations of motion:

$$\dot{r}_p = -w \cos\phi\ , \qquad w > 1,$$

$$\dot{r}_e = w \cos\psi,$$

$$\dot{\theta} = \frac{\sin\psi}{r_e} - w\frac{\sin\phi}{r_p}.$$

Player P chooses ϕ and player E chooses ψ.

The Outcome Functional:

The payoff is:

$$G(r_p, r_e, \theta) = 0 \quad , \quad L(r_p, r_e, \theta) = 1$$

This is a pure pursuit evasion game. Player P wants to terminate and wishes to minimize the outcome and player E wishes to maximize the outcome.

1. Solve for the candidate optimal trajectories emanating backwards from the target-set. (Show that J_θ is constant along each optimal trajectory).

2. Assume $J_\theta = \beta/(w-1)$ where β is a parameter.
Show that the optimal trajectories are straight lines.
Show that the trajectory for $\beta = 1$ is tangent to the obstacle disc.

3. Rename r_p as x and r_e as y and use another definition for the controls:
Player P chooses b such that $|b| \leq x$ where $\frac{b}{x} = \sin\phi$,
Player E chooses c such that $|c| \leq y$ where $\frac{c}{y} = \sin\psi$.
redo 1. and 2.

•

Problem 7.5.7

Consider the *linear quadratic differential game* (see *T. Basar* [Bas90]):

The game-set:

$$\mathcal{S} \equiv \{(x,t) \mid \ x \in R^1 \ , \ 0 \leq t \leq t_f\}.$$

The target-set:

$$\mathcal{C} \equiv \{(x,t) \mid \ x \in R^1 \ , \ t = t_f\}.$$

The equations of motion:

$$\dot{x} = u + v,$$

$$\dot{t} = 1.$$

The outcome functional:

$$\pi[x, t, u(\cdot), v(\cdot)] = \int_t^{t_f} (x^2 + u^2 - \gamma^2 v^2) dt.$$

1. Find the optimal controls of players P and E depending on the parameters (γ, t_f).

2. Find the range of the parameter γ for which solution for the *infinite horizon* version of the game ($t_f \to \infty$) exists.

3. Define the *controller design problems* that corresponds to previous parts and find the infimum of the achievable attenuation for each case.

•

Problem 7.5.8 (LQG with norm bounded controls)

Consider the following linear quadratic differential game (see *S. Gutman* and *G. Leitmann* in [GL75]).

The game-set:

$$\mathcal{S} \equiv \{(x,t) \mid \ x \in R^n \ , \ t_0 \le t \le t_1\}.$$

The target-set:

$$\mathcal{C} \equiv \{(x,t) \mid \ x \in R^n \ , \ t = t_1\}.$$

The equations of motion:

$$\dot{x} = A(t)x + b(u+v) \ , \quad x(t_0) = x_0,$$

$$\dot{t} = 1.$$

$A(\cdot)$ is an $n \times n$ continuous matrix function, b is a constant $n \times 1$ vector.
At each instance player P chooses u and player E chooses v:

$$u(t) \in \mathcal{U} \ , \quad \mathcal{U} \equiv \{u \mid |u| \le \rho_u\},$$

$$v(t) \in \mathcal{V} \ , \quad \mathcal{V} \equiv \{v \mid |v| \le \rho_v\}.$$

The outcome functional:

$$\pi[x,t,u(\cdot),v(\cdot)] = \frac{1}{2}\int_t^{t_1} x^T Q x dt,$$

$Q(\cdot)$ is a symmetric continuous matrix function.

1. Assume that a candidate *value function* is for $J_x b \geq 0$:

$$J(x,t) = \frac{1}{2}x^T P x - x^T Rb(\rho_u - \rho_v) + (\rho_u - \rho_v)^2 \int_t^{t_1} b^T Rbds,$$

and for $J_x b \leq 0$

$$J(x,t) = \frac{1}{2}x^T P x + x^T Rb(\rho_u - \rho_v) + (\rho_u - \rho_v)^2 \int_t^{t_1} b^T Rbds.$$

Show that it is necessary that P and R should satisfy the following equations:

$$\begin{aligned} \dot{P} + PA + A^T P + Q = 0 \;, \quad & P(t_1) = 0, \\ \dot{R} + A^T R + P = 0 \;, \quad & R(t_1) = 0. \end{aligned} \tag{7.101}$$

•

8

Construction of Semipermeable Surfaces

8.1 Introduction

Semipermeable surfaces (SPS) appear often in differential games.
We have already seen two kinds of surfaces that are necessarily semipermeable: the *isovalue surfaces* in a *terminal cost* differential game (see Corollary 5.5.2) and the boundary of the *escape-set* F (see section 5.3). Later we shall see that even *open barriers*, (surfaces across which the value function $J(\mathbf{x})$ suffers a jump-discontinuity), are necessarily semipermeable surfaces.

If we examine the definition of *semipermeability* (see chapter 4) we observe that it depends only on the properties of the *equations of motion* $\dot{\mathbf{x}} = \mathbf{f}(\mathbf{x}, \mathbf{u}, \mathbf{v})$ and that it does not involve the notions of the *target-set* and the *outcome functional*.
The semipermeablity of a surface in the context of a certain differential game is hence not particular only to this game but is a more general property in the sense that it is related to a whole class of differential games that have common equations of motion.

*R. Isaacs*was first to deduce a procedure for constructing semipermeable surfaces, based on the techniques of constructing candidate optimal trajectories in differential games.

Let us consider a *terminal cost differential game* with a separable *velocity vector* $\mathbf{f}(\mathbf{x}, \mathbf{u}, \mathbf{v})$.
The corresponding *Isaacs equations* are:

$$\min_{\mathbf{u}\in\mathcal{U}} \max_{\mathbf{v}\in\mathcal{V}} \nabla J(\mathbf{x}) \cdot \mathbf{f}(\mathbf{x}, \mathbf{u}, \mathbf{v}) = \mathbf{0}. \tag{8.1}$$

Deviding relation (8.1) by the constant $|\nabla J(\mathbf{x})|$ and recalling that $\frac{\nabla J(\mathbf{x})}{|\nabla J(\mathbf{x})|}$ is the unit normal $\mathbf{n}(\mathbf{x})$ to the isovalue surface we get[1] that:

$$\min_{\mathbf{u}\in\mathcal{U}} \max_{\mathbf{v}\in\mathcal{V}} \mathbf{n}(\mathbf{x}) \cdot \mathbf{f}(\mathbf{x}, \mathbf{u}, \mathbf{v}) = \mathbf{0}. \tag{8.2}$$

---

[1]See Corollary 5.5.2.

R. Isaacs noted that relation (8.2) represents relations ME1 and ME2 for $\mathbf{n}(\mathbf{x})$ just as relation (8.1) represents relations ME1 and ME2 for $\nabla J(\mathbf{x})$.
He also noted that the *barrier paths* obtained by integrating $\dot{\mathbf{x}} = \mathbf{f}(\mathbf{x}, \overset{*}{\mathbf{u}}, \overset{*}{\mathbf{v}})$ (where $(\overset{*}{\mathbf{u}}, \overset{*}{\mathbf{v}})$ solve relation (8.2)) are analogous to the optimal trajectories corresponding to relation (8.1).
He carried this analogy further and found (as could easily be verified by the reader) that a derivation similar to that given in section 5.6 provide that along *barrier paths* the following *adjoint equations* hold:

$$\overset{\circ}{\mathbf{n}}(\mathbf{x}) = \frac{\partial H(\mathbf{x}, \mathbf{n}, \overset{*}{\mathbf{u}}, \overset{*}{\mathbf{v}})}{\partial x}. \tag{8.3}$$

R. Isaacs noted that only (n-1) independent components are needed to specify the unit normal $n(\mathbf{x})$.
He concluded that a procedure, almost similar to that of section 7.2, can be used for constructing manifolds of *barrier paths* emanating backwards from a given (n-2) dimensional curve R.

It is indeed remarkable that the procedures for constructing candidate optimal trajectories in differential games led to procedures for constructing regular semipermeable surfaces (called *natural barriers* by Isaacs) which have, as we have mentioned, a much more general significance.

8.2 Construction of Semipermeable Surfaces

8.2.1 The Regular Construction

It is often possible to construct a manifold B of *barrier paths* emanating from a given (n-2)-dimensional curve R by a method that resembles that of section 7.2.
In order to initialize such a procedure initial values of (n-1) components of $n(\mathbf{x})$ at points $\mathbf{x}$ on the curve R are required.
Let us see how this can actually be done.

Let us parametrize the state $\mathbf{x}$ on R as $\mathbf{x} = \mathbf{h}(\mathbf{l})$ where $\mathbf{l}$ is a (n-2)-dimensional vector:

$$x_i = h_i(l_1, l_2, \cdots, l_{n-2}).$$

Barrier paths reach R at $\tau = 0$ (τ is *retro-time*).
A barrier path that emanates from $\mathbf{h}(\mathbf{l})$ is given by:

$$\mathbf{y}(\mathbf{l}, \tau) = \mathbf{h}(\mathbf{l}) - \int_0^\tau \overset{*}{\mathbf{f}}\,[\mathbf{y}(\mathbf{l}, \alpha)]d\alpha, \tag{8.4}$$

where

$$\overset{*}{\mathbf{f}}\,[\mathbf{y}(\mathbf{l},\tau)] = \mathbf{f}(\mathbf{y}, \overset{*}{\mathbf{u}}\,(\mathbf{y}), \overset{*}{\mathbf{v}}\,(\mathbf{y})). \tag{8.5}$$

Obviously

$$\frac{\partial \mathbf{y}(\mathbf{l},\tau)}{\partial \tau} = -\,\overset{*}{\mathbf{f}}\,[\mathbf{y}(\mathbf{l},\tau)]. \tag{8.6}$$

The adjoint equations are

$$\frac{\partial \mathbf{n}[\mathbf{y}(\mathbf{l},\tau)]}{\partial \tau} = \frac{\partial H\{[\mathbf{y}(\mathbf{l},\tau)], \mathbf{n}[\mathbf{y}(\mathbf{l},\tau)], \overset{*}{\mathbf{u}}\,[\mathbf{y}(\mathbf{l},\tau)], \overset{*}{\mathbf{v}}\,[\mathbf{y}(\mathbf{l},\tau)]\}}{\partial \mathbf{y}}. \tag{8.7}$$

Obviously

$$\mathbf{n}[\mathbf{y}(\mathbf{l},0)] = \mathbf{n}[\mathbf{h}(\mathbf{l})].$$

In order to initialize the integration of relations (8.7) we need to have the values of $n_i[\mathbf{h}(\mathbf{l})]$.

Consider a point $\mathbf{x} \in R$. Let $d\mathbf{x}$ be any infinitesimal displacement vector such that also $\mathbf{x} + d\mathbf{x} \in R$. The components dx_i of $d\mathbf{x}$ are given by:

$$dx_i = \sum_{j=1}^{n-2} \frac{\partial h_i(\mathbf{l})}{\partial l_j} dl_j. \tag{8.8}$$

The displacement $d\mathbf{x}$ lies also on the surface B that we want to construct (since obviously $R \in B$) and is orthogonal to the *normal* $\mathbf{n}(\mathbf{x})$ to B at $\mathbf{x}$.

$$\sum_{i=1}^{n} n_i(\mathbf{x}) \cdot dx_i = 0, \tag{8.9}$$

$$\sum_{i=1}^{n} n_i[\mathbf{h}(\mathbf{l})] \sum_{j=1}^{n-2} \frac{\partial h_i(\mathbf{l})}{\partial l_j} dl_j = 0. \tag{8.10}$$

We can simplify relation (8.10) by the use of a specific set of (n-2) independent displacements $d\mathbf{x}$ that span R

$$\begin{bmatrix} dl_1 \\ 0 \\ \cdot \\ \cdot \\ \cdot \\ \cdot \\ \cdot \\ 0 \end{bmatrix} \begin{bmatrix} 0 \\ dl_2 \\ \cdot \\ \cdot \\ \cdot \\ \cdot \\ \cdot \\ 0 \end{bmatrix} \cdots \begin{bmatrix} 0 \\ \cdot \\ \cdot \\ \cdot \\ dl_m \\ \cdot \\ \cdot \\ 0 \end{bmatrix} \cdots \begin{bmatrix} 0 \\ \cdot \\ \cdot \\ \cdot \\ \cdot \\ \cdot \\ \cdot \\ dl_{n-2} \end{bmatrix}. \tag{8.11}$$

For each displacement that belong to this set relation (8.10) now becomes:

$$\sum_{i=1}^{n} n_i[\mathbf{h}(\mathbf{l})] \frac{\partial h_i(\mathbf{l})}{\partial l_m} dl_m = 0, \tag{8.12}$$

and since relation (8.12) must hold for arbitrary magnitude of dl_m it can be expressed in a differential form as

$$\sum_{i=1}^{n} n_i[\mathbf{h}(\mathbf{l})]\frac{\partial h_i(\mathbf{l})}{\partial l_m} = 0. \tag{8.13}$$

equation (8.13) yields (n-2) independent equations between the (n-1) required $n_i[\mathbf{h}(\mathbf{l})]$.
The additional relation is supplied by specializing *ME2* to points on R:

$$\sum_{i=1}^{n} n_i[\mathbf{h}(\mathbf{l})] \cdot f_i\{\mathbf{h}(\mathbf{l}), \overset{*}{\mathbf{u}}[\mathbf{h}(\mathbf{l}], \overset{*}{\mathbf{v}}[\mathbf{h}(\mathbf{l})]\} = 0. \tag{8.14}$$

Notice that only when $\mathbf{f}\{\mathbf{h}(\mathbf{l}), \overset{*}{\mathbf{u}}[\mathbf{h}(\mathbf{l}], \overset{*}{\mathbf{v}}[\mathbf{h}(\mathbf{l})]\}$ is not tangent to the curve R at $\mathbf{x}$ relation (8.14) is independent of relations (8.13) and we can use them to to obtain $n_i[\mathbf{h}(\mathbf{l})]$.

8.2.2 Semipermeability of the Constructed Manifold

Thus far we have described a procedure for constructing a (n-1)-dimensional manifold B of *candidate barrier paths* through a given (n-2)-dimensional curve R.
Let us now show that B is indeed a semipermeable surface[2].

To do so we shall again use a parametrization $\mathbf{x} = \bar{\mathbf{h}}(\bar{\mathbf{l}})$ of the state $\mathbf{x} \in B$ with respect to a vector variable $\bar{\mathbf{l}}$ obtained by adjoining a component $l_{n-1} = \tau$ as an additional component to the components of the vector $\mathbf{l}$ defined previously.

$$\bar{\mathbf{l}} = (\bar{l}_1, \cdots, \bar{l}_m, \cdots, \bar{l}_{n-1}),$$

where

$$\bar{l}_m = l_m \quad , \text{ for } m = 1, \cdots, (n-2),$$

$$\bar{l}_{n-1} = \tau.$$

Obviously:

$$\begin{aligned} \bar{\mathbf{h}}(\bar{\mathbf{l}}) &= \mathbf{y}(\mathbf{l}, \tau), \\ \mathbf{h}(\mathbf{l}) &= \mathbf{y}(\mathbf{l}, 0), \end{aligned} \tag{8.15}$$

and

$$\frac{\partial \bar{\mathbf{h}}(\bar{\mathbf{l}})}{\partial \tau} = -\mathbf{f}\{\bar{\mathbf{h}}(\bar{\mathbf{l}}), \overset{*}{\mathbf{u}}[\bar{\mathbf{h}}(\bar{\mathbf{l}})], \overset{*}{\mathbf{v}}[\bar{\mathbf{h}}(\bar{\mathbf{l}})]\}. \tag{8.16}$$

---

[2]We follow the proof given by *R. Isaacs* in [Isa75]

Consider a point $\mathbf{x} \in B$. Any displacement $d\mathbf{x}$ (with components dx_i) such that $\mathbf{x} + d\mathbf{x} \in B$ can be expressed as:

$$dx_i = \sum_{j=1}^{n-1} \frac{\bar{h}_i(\bar{\mathbf{l}})}{\partial \bar{l}_j} d\bar{l}_j. \tag{8.17}$$

We shall first prove that the constructed candidate $\mathbf{n}(\mathbf{x})$ is indeed normal to the manifold B at $\mathbf{x}$. We have to prove that

$$\mathbf{n}(\mathbf{x}) \cdot d\mathbf{x} = 0.$$

We have that:

$$\mathbf{n}(\mathbf{x}) \cdot d\mathbf{x} = \sum_{i=1}^{n} n_i[\bar{\mathbf{h}}(\bar{\mathbf{l}})] \sum_{j=1}^{n-1} \frac{\bar{h}_i(\bar{\mathbf{l}})}{\partial \bar{l}_j} d\bar{l}_j. \tag{8.18}$$

Relation (8.18) can be simplified considerably if we choose the following specific set of (n-1) vectors $d\bar{\mathbf{l}}$,

$$\begin{bmatrix} d\bar{l}_1 \\ 0 \\ \cdot \\ \cdot \\ \cdot \\ \cdot \\ \cdot \\ 0 \end{bmatrix} \cdots \begin{bmatrix} 0 \\ \cdot \\ \cdot \\ \cdot \\ d\bar{l}_m \\ \cdot \\ \cdot \\ 0 \end{bmatrix} \cdots \begin{bmatrix} 0 \\ \cdot \\ \cdot \\ \cdot \\ \cdot \\ \cdot \\ d\bar{l}_{n-2} \\ 0 \end{bmatrix} \cdots \begin{bmatrix} 0 \\ \cdot \\ \cdot \\ \cdot \\ \cdot \\ \cdot \\ 0 \\ d\bar{l}_{n-1} \end{bmatrix}. \tag{8.19}$$

Relation (8.18) now becomes:

$$\mathbf{n}(\mathbf{x}) \cdot d\mathbf{x} = \sum_{i=1}^{n} n_i[\bar{\mathbf{h}}(\bar{\mathbf{l}})] \frac{\partial \bar{h}_i(\bar{\mathbf{l}})}{\partial \bar{l}_m} d\bar{l}_m. \tag{8.20}$$

Let us define $Q_m(\bar{\mathbf{l}})$ as:

$$Q_m(\bar{\mathbf{l}}) = \sum_{i=1}^{n} n_i[\bar{\mathbf{h}}(\bar{\mathbf{l}})] \frac{\partial \bar{h}_i(\bar{\mathbf{l}})}{\partial \bar{l}_m}. \tag{8.21}$$

In order to show that $\mathbf{n}(\mathbf{x}) \cdot d\mathbf{x} = 0$ for any displacement $d\mathbf{x}$ such that $\mathbf{x} + d\mathbf{x} \in B$ it suffices to show that $Q_m(\bar{\mathbf{l}}) \equiv 0$ for all m and $\bar{\mathbf{l}}$.

Lemma 8.2.1

$$Q_{n-1}(\bar{l}_1, \cdots, \bar{l}_{n-2}, \bar{l}_{n-1}) = 0.$$

Proof

Recalling that $\bar{l}_{n-1} = \tau$ we have from relation (8.16)

$$Q_{n-1}(\bar{l}_1, \cdots, \bar{l}_{n-2}, \bar{l}_{n-1}) = -\sum_{i=1}^{n} n_i[\bar{\mathbf{h}}(\bar{\mathbf{l}})] \cdot f_i\{\bar{\mathbf{h}}(\bar{\mathbf{l}}), \overset{*}{\mathbf{u}}[\bar{\mathbf{h}}(\bar{\mathbf{l}}], \overset{*}{\mathbf{v}}[\bar{\mathbf{h}}(\bar{\mathbf{l}})]\}. \tag{8.22}$$

In the construction procedure we chose *barrier controls* $(\overset{*}{\mathbf{u}}, \overset{*}{\mathbf{v}})$ that satisfy ME2 so evidently the right hand side of relation (8.22) vanishes along the constructed *barrier paths*.

•

Lemma 8.2.2

$$Q_m(\bar{l}_1, \cdots, \bar{l}_{n-2}, 0) = 0,$$

for $m = 1, \cdots, (n-2)$.

Proof

For $m = 1, \cdots, (n-2)$ we substitute from relation (8.15) into relation (8.21) and get

$$Q_m(\bar{l}_1, \cdots, \bar{l}_{n-2}, 0) = \sum_{i=1}^{n} n_i[\mathbf{h}(\mathbf{l})] \frac{\partial h_i(\mathbf{l})}{\partial \bar{l}_m}. \tag{8.23}$$

In the construction of the *barrier paths* we chose $n_i[\mathbf{h}(\mathbf{l})]$ that satisfy relation (8.12) so evidently the right hand side of relation (8.23) equals 0.

•

Lemma 8.2.3

$$\frac{\partial Q_m(\bar{l}_1, \cdots, \bar{l}_{n-1})}{\partial \tau} = 0, \tag{8.24}$$

for $m = 1, \cdots, (n-1)$.

Proof

Differentiating relation (8.21) with respect to τ we obtain:

$$\frac{\partial Q_m(\bar{\mathbf{l}})}{\partial \tau} = \sum_{i=1}^{n} \frac{\partial n_i[\bar{\mathbf{h}}(\bar{\mathbf{l}})]}{\partial \tau} \frac{\partial \bar{h}_i(\bar{\mathbf{l}})}{\partial \bar{l}_m} - \sum_{i=1}^{n} n_i[\bar{\mathbf{h}}(\bar{\mathbf{l}})] \frac{\partial \overset{*}{f}_i [\bar{\mathbf{h}}(\bar{\mathbf{l}})]}{\partial \bar{l}_m}. \tag{8.25}$$

Evaluating $\frac{\partial \overset{*}{f}_i[\bar{\mathbf{h}}(\bar{\mathbf{l}})]}{\partial \bar{l}_m}$ we get:

$$\frac{\partial Q_m(\bar{\mathbf{l}})}{\partial \tau} = \sum_{i=1}^{n} \frac{\partial n_i[\bar{\mathbf{h}}(\bar{\mathbf{l}})]}{\partial \tau} \frac{\partial \bar{h}_i(\bar{\mathbf{l}})}{\partial \bar{l}_m} - \sum_{i=1}^{n} n_i[\bar{\mathbf{h}}(\bar{\mathbf{l}})]\{\sum_{j=1}^{n} \frac{\partial \overset{*}{f}_i[\bar{\mathbf{h}}(\bar{\mathbf{l}})]}{\partial \bar{h}_j(\bar{\mathbf{l}})} \frac{\partial \bar{h}_j(\bar{\mathbf{l}})}{\partial \bar{l}_m} +$$
$$\sum_{j=1}^{n} \sum_{k=1}^{p} \frac{\partial \overset{*}{f}_i[\bar{\mathbf{h}}(\bar{\mathbf{l}})]}{\partial \overset{*}{u}_k} \frac{\partial \overset{*}{u}_k}{\partial \bar{h}_j(\bar{\mathbf{l}})} \frac{\partial \bar{h}_j(\bar{\mathbf{l}})}{\partial \bar{l}_m} + \sum_{j=1}^{n} \sum_{k=1}^{q} \frac{\partial \overset{*}{f}_i[\bar{\mathbf{h}}(\bar{\mathbf{l}})]}{\partial \overset{*}{v}_k} \frac{\partial \overset{*}{v}_k}{\partial \bar{h}_j(\bar{\mathbf{l}})} \frac{\partial \bar{h}_j(\bar{\mathbf{l}})}{\partial \bar{l}_m}\}. \tag{8.26}$$

Since $\overset{*}{\mathbf{u}}$ and $\overset{*}{\mathbf{v}}$ are solutions of the Isaacs *ME1* and *ME2* equations we can apply the result of equation 5.34 with $d\mathbf{x}$ that corresponds to each of the base vectors $d\bar{\mathbf{l}} = (0, \cdots, \bar{l}_m, \cdot, 0)$ and get:

$$\sum_{i=1}^{n} \sum_{j=1}^{n} \sum_{k=1}^{p} n_i[\bar{\mathbf{h}}(\bar{\mathbf{l}})] \frac{\partial \overset{*}{f}_i [\bar{\mathbf{h}}(\bar{\mathbf{l}})]}{\partial \overset{*}{u}_k} \frac{\partial \overset{*}{u}_k}{\partial \bar{h}_j(\bar{\mathbf{l}})} \frac{\partial \bar{h}_j(\bar{\mathbf{l}})}{\partial \bar{l}_m} d\bar{l}_m = H_{\overset{*}{u}}\, \overset{*}{\mathbf{u}}_x\, d\mathbf{x} = 0, \tag{8.27}$$

$$\sum_{i=1}^{n}\sum_{j=1}^{n}\sum_{k=1}^{q} n_i[\bar{\mathbf{h}}(\bar{\mathbf{l}})]\frac{\partial \overset{*}{f_i}\,[\bar{\mathbf{h}}(\bar{\mathbf{l}})]}{\partial \overset{*}{v}_k}\frac{\partial \overset{*}{v}_k}{\partial \bar{h}_j(\bar{\mathbf{l}})}\frac{\partial \bar{h}_j(\bar{\mathbf{l}})}{\partial \bar{l}_m}d\bar{l}_m = H_{\overset{*}{v}}\ \overset{*}{\mathbf{v}}_x\ d\mathbf{x} = 0. \quad (8.28)$$

The adjoint equations provide that:

$$\frac{\partial n_i[\bar{\mathbf{h}}(\bar{\mathbf{l}})]}{\partial \tau} = \frac{\partial H\{\bar{\mathbf{h}}(\bar{\mathbf{l}}), \mathbf{n}[\bar{\mathbf{h}}(\bar{\mathbf{l}})], \overset{*}{\mathbf{u}}, \overset{*}{\mathbf{v}}\}}{\partial \bar{h}_i(\bar{\mathbf{l}})} = \sum_{j=1}^{n} n_j[\bar{\mathbf{h}}(\bar{\mathbf{l}})]\frac{\partial \overset{*}{f_j}\,[\bar{\mathbf{h}}(\bar{\mathbf{l}})]}{\partial \bar{h}_i(\bar{\mathbf{l}})}, \quad (8.29)$$

so

$$\sum_{i=1}^{n}\frac{\partial n_i[\bar{\mathbf{h}}(\bar{\mathbf{l}})]}{\partial \tau}\frac{\partial \bar{h}_i(\bar{\mathbf{l}})}{\partial \bar{l}_m} = \sum_{i=1}^{n}\sum_{j=1}^{n} n_j[\bar{\mathbf{h}}(\bar{\mathbf{l}})]\frac{\partial \overset{*}{f_j}\,[\bar{\mathbf{h}}(\bar{\mathbf{l}})]}{\partial \bar{h}_i(\bar{\mathbf{l}})}\frac{\partial \bar{h}_i(\bar{\mathbf{l}})}{\partial \bar{l}_m}. \quad (8.30)$$

Interchanging the dummy variables i and j in the right hand side of equation (10.6.1) and substituting into equation (8.26) completes the proof.

•

Lemma 8.2.4

Let B be a manifold of *barrier paths* constructed by the method of section 8.2 then at each $\mathbf{x} \in B$

$$\sum_{i=1}^{n} n_i[\bar{\mathbf{h}}(\bar{\mathbf{l}})] f_i\{\bar{\mathbf{h}}(\bar{\mathbf{l}}), \overset{*}{\mathbf{u}}\,[\bar{\mathbf{h}}(\bar{\mathbf{l}})], \overset{*}{\mathbf{v}}\,[\bar{\mathbf{h}}(\bar{\mathbf{l}})]\} = 0, \quad (8.31)$$

and B is semipermable.

Proof
The proof follows from the previous lemmas and from the construction and is left to the reader.

•

Remarks.

1. We must remember that the validity of the adjoint equations depends on the validity of the assumption that the surface B is *smooth* ($\mathbf{n}(\mathbf{x})$ is continuous on B). An *ill* behavior of the *barrier paths* (as in a case when $\mathbf{n}(\mathbf{x})$ is nonunique) indicates that the smoothness assumption is not valid any more.
2. Semipermeabilty of a surface B at a point $\mathbf{x}$ is a *local* property of the surface while the *saddle property* of the value $J(\mathbf{x})$ is related to a *global* property of the solution of a differential game.

For this reason the semipermeability of manifolds of constructed barrier paths is inherently guaranteed[3] *Barrier Surfaces, composite* while the validity of constructed candidate optimal trajectories is not guaranteed by the construction procedure itself and extra global conditions have to be met.

8.3 Examples

Example 8.3.1 (The Dolichobrachistochrone) *[Isa75,Chi76]*

Let us construct semipermeable surfaces for the *Dolichobrachistochrone* problem (see Example 5.2.1).

The game-set:

The game takes place in a quadrant of a plane.

$$\mathcal{S} \equiv \{(x,y) \mid x \geq 0,\ y \geq 0\}\,.$$

The target-set:

$$\mathcal{C} \equiv \{(x,y) \mid x = 0,\ y \geq 0\}\,.$$

The equations of motion:

$$\dot{x} = \sqrt{y}\cos\varphi + \tfrac{1}{2}w(\psi+1) \quad , \ |\psi| \leq 1,$$

$$\dot{y} = \sqrt{y}\sin\varphi + \tfrac{1}{2}w(\psi-1).$$

Player P chooses φ and player E chooses ψ.

This is a 2-dimensional game so we shall try to construct *semipermeable surfaces* through *points* on the y-axis.

The parametrization

A point is 0-dimensional so parametrization is not relevant here.

The barrier controls

The equations of motion are separable so we can write ME1 as:

$$\min_{\varphi}\max_{|\psi|\leq 1}\{n_x[\sqrt{y}\cos\varphi + \frac{1}{2}w(\psi+1)] + n_y[\sqrt{y}\sin\varphi + \frac{1}{2}w(\psi-1)]\}. \quad (8.32)$$

Let us define $A(x,y)$ as

$$A(x,y) = n_x(x,y) + n_y(x,y).$$

[3] This applies to the *regular* parts of the *barrier*. Corners *corner* of *composite barriers* deserve a different analysis.

The optimal *barrier control* for player E is:

$$\overset{*}{\psi}(x,y) = \text{sign}A(x,y)\ , \qquad \text{for } A(x,y) \neq 0,$$
$$\overset{*}{\psi}(x,y) = \text{ any admissible } \psi\ , \quad \text{for } A(x,y) = 0. \tag{8.33}$$

The *barrier control* for player P is:

$$(\sin \overset{*}{\varphi}(x,y), \cos \overset{*}{\varphi}(x,y)) \parallel (-n_x(x,y), -n_y(x,y)). \tag{8.34}$$

The adjoint equations

We shall assume that adjacent to the y-axis $\text{sign}[n_x(x,y) + n_y(x,y)] = -1$ (representing the requirement that the barrier paths will emanate into the game-set).

$$\overset{\circ}{n}_x [x(\tau), y(\tau)] = 0,$$
$$\overset{\circ}{n}_y [x(\tau), y(\tau)] = \tfrac{\sigma}{2\sqrt{y}}. \tag{8.35}$$

where $\sigma = \sqrt{n_x^2 + n_y^2}$.

We also have

$$\overset{\circ}{A} [x(\tau), y(\tau)] = \frac{\sigma}{2\sqrt{y}} > 0. \tag{8.36}$$

We can now conclude that indeed our assumption is true and that adjacent to the target-set $A(x,y) < 0$ so that barrier paths arrive at the target-set with $\overset{*}{\psi} = -1$.

Solving the adjoint equation for $n_x[x(\tau), y(\tau)]$ we get[4]:

$$n_x[x(\tau), y(\tau)] = 1. \tag{8.37}$$

ME2 is:

$$-\sqrt{y}\sqrt{n_x^2 + n_y^2} + \frac{1}{2}w|n_x + n_x| + \frac{1}{2}w[n_x + n_x] = 0. \tag{8.38}$$

By using ME2 we can obtain $n_y[x(\tau), y(\tau)]$

$$n_y[x(\tau), y(\tau)] = -\sqrt{y(\tau)/[w^2 - y(\tau)]}. \tag{8.39}$$

ME1 is homogeneous in n_x and n_y so we can substitue $-\cos \overset{*}{\varphi}$ for n_x and $-\sin \overset{*}{\varphi}$ for n_y and get:

$$-\cos \overset{*}{\varphi} (\sqrt{y} \cos \overset{*}{\varphi}) - \sin \overset{*}{\varphi} (\sqrt{y} \sin \overset{*}{\varphi} - w) = 0. \tag{8.40}$$

---

[4] We assume $n_x[x(0), y(0)] = 1$. This choice is arbitrary because the direction of the normal is not changed if all its components are increased proportionally.

We find that:

$$\sin \overset{*}{\varphi} = \frac{\sqrt{y}}{w}. \tag{8.41}$$

This means that the construction is valid only for $y \leq w^2$.
We can obtain a differential equation for the barrier paths:

$$\frac{dy}{dx} = \frac{\dot{y}}{\dot{x}} = -\frac{n_x}{n_y} = -\frac{1}{\tan \overset{*}{\varphi}} = -\sqrt{\frac{w^2 - y}{y}}, \tag{8.42}$$

so that for $y \leq w^2$ the equation of the barrier paths is:

$$x = -\sqrt{y(w^2 - y)} + \frac{1}{2} w^2 \sin^{-1} \frac{2y - w^2}{w^2} + c. \tag{8.43}$$

•

8.4 Problems

Problem 8.4.1 (The Dolichobrachistochrone)

Consider Example 8.3.1.

1. Show that the corners between the optimal trajectory that emanates from $(0, w^2/2)$ and the surfaces $y = w^2$ and $x = 0$ do not leak.

•

Problem 8.4.2 (The Homicidal Chauffeur)

Consider the *Homicidal Chauffeur* problem (defined in section 13.3).

1. Construct a semipermeable surface emanating into the game-set from the point $(r = \beta,\ \theta_{BUP} = \cos^{-1} \gamma)$ and show that it is given by

$$x(\tau) = 1 - \cos \tau + (\beta - \gamma\tau) \sin(\theta_{BUP} + \tau),$$

$$y(\tau) = \sin \tau + (\beta - \gamma\tau) \cos(\theta_{BUP} + \tau).$$

2. Show that a construction of such a surface is feasible only for

$$\beta^2 + \gamma^2 \leq 1.$$

•

9

A Topography of the Map of Isovalue-Surfaces

A preview on singular surfaces.

9.1 Introduction

In a differential game we allow the value function $J(\mathbf{x})$ to have a jump discontinuity across certain surfaces B in the game space. We call such surfaces *barriers* . We also allow $\nabla J(\mathbf{x})$, the gradient of the value function, to have a jump discontinuity across certain surfaces D in the game space. We often call such surfaces *singular surfaces*.
We can draw an analogy between a *topographic map* and a map of *iso-value surfaces*, (surfaces where $J(\mathbf{x}) = c$). *Barriers* represent *cliffs* and some *singular surfaces* represent *valleys* and *mountain ridges.*
We shall see that optimal trajectories, just like roads, do not cross *barriers* but may go through or follow, *singular surfaces.*
In this chapter we give a preliminary description of various types of surfaces that may appear in a map of iso-value surfaces. We use the terminology of *R. Isaacs* and *J. V. Breakwell.*

9.2 Barriers and Safe Contact

9.2.1 Barriers

A *barrier* is a surface at which the value function is discontinuous. A barrier is never crossed by optimal trajectories.
We distinguish between *open barriers* and *closed barriers*. If the function $J(\mathbf{x})$ has a finite jump-discontinuity at the barrier then it is an *open barrier* and it belongs to the interior of the *capture-set* $\mathcal{W}$. If $J(\mathbf{x})$ is defined only at one side of the barrier then it is a *closed barrier* and it belongs to the boundaries of the capture-set. *Open barriers* are analogous to *cliffs* and *closed barriers* are analogous to *canyons.*
Barriers play an important role in solving conflict problems. The boundary that separates a *capture-set* $\mathcal{W}$ from the *escape-set* $\mathcal{F}$ is a *closed barrier.*

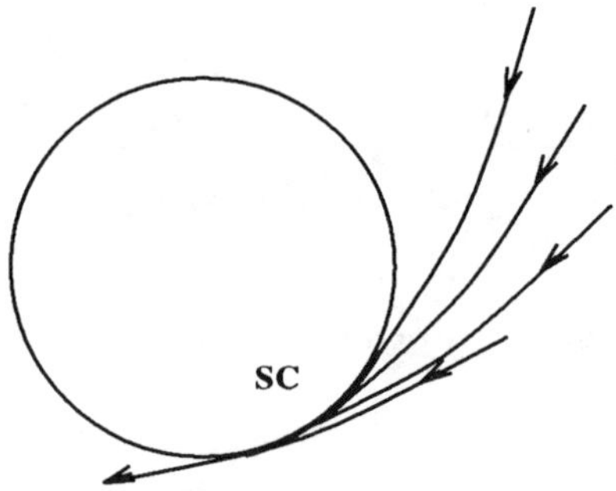

FIGURE 9.1. Safe contact with a tangential junction

9.2.2 State Costraints

A surface that belongs to the boundary of the game-set and is not considered as part of the target-set is called a *state constraint*. Both players are not allowed to cross a *state constraint*.
We saw a *state costraint* in Example 2.7.1 where both players where not allowed to cross the *obstacle*.

9.2.3 Safe Contact

In many problems we find that some optimal trajectories have a *singular arc* segment that is adjacent to a barrier or even to the target-set! We sometimes call this segment *safe contact (SC)*.
Such a phenomenon is no surprise in practice. The majority of the important battles in history where fought in the surrounding of a topographic obstacle where small mistakes by one side may mean a great change in the result. This is also typical to a *brinkmanship* policy which risks great change in the state of affairs.

9.2.4 The Tributaries

If we look on the map of optimal trajectories we sometimes see regular trajectories arrive at a barrier (or at the target-set) proceed as a safe-contact singular arc and leave the barrier (or the target-set) to become regular trajectories onward. The picture of regular optimal trajectories joining singular arcs resembles the situation in nature where small tributary creeks join bigger streams and rivers. This is the reason why we sometimes call the field of optimal trajectories emmanating backward from a singular surface *the field of tributaries* to the singular surface.
We distinguish between two types of junctions of tributary optimal trajec-

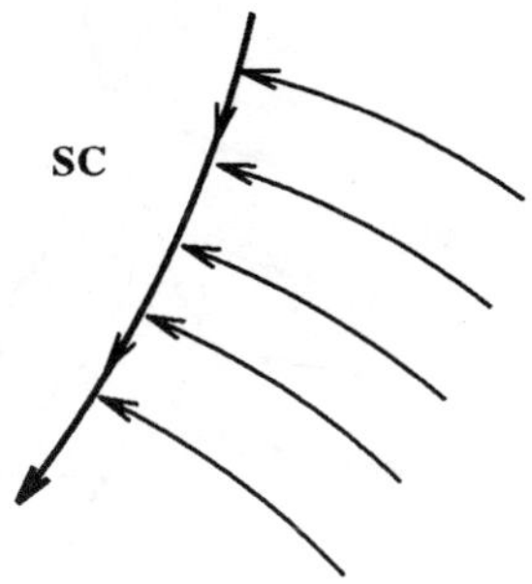

FIGURE 9.2. Safe contact with a transversal junction

tories to singular arcs:

Tangential junction
The regular arc joins the singular arc smoothly. (See Figure 9.1).

Transverse junction
The regular arc joins the singular arc transversely. (See Figure 9.2).

9.3 Switch Surfaces

In a topographic map we find certain lines that do not represent the geography but refer to it. Lines that demark the time regions are such. When we cross such a line we have to switch the hour in our watches.
In the map of iso-value surfaces we find *switch surfaces*. They do not relate to the *value* but they mark a switch of controls (of the *bang-bang* type) by one of the players as a trajectory crosses them.

9.4 Dispersal Surfaces

A *dispersal surface (DS)* is analogous to a ridge that constitutes a boundary between two river water systems. From a point on such a ridge water can flow either ways but once it flows it remains in that side.
A dispersal surface is a locus of points from which optimal trajectories *leave to both sides.* (See Figure 9.3).
$\nabla J(\mathbf{x})$ is discontinuous across a dispersal surface but singular arcs of optimal trajectories do not adhere to it.
If the state of a game is at a point on a dispersal surface it means that one player has two equally good choices for his control. If he chooses one the

FIGURE 9.3. A dispersal surface

trajectory may go to one side of the dispersal surface and if he chooses the other the optimal trajectory will take the other direction. It is a momentary choice, like a choice made at a cross road, and the decision cannot be delayed.

9.5 Universal and Focal Surfaces

9.5.1 General characterization

Universal and Focal surfaces are analogous to river beds. Creeks of water join from each side and merge into the main stream. Optimal trajectories *arrive* at those surfaces from both sides. The difference between the two is in the nature of the junction of their tributaries.

9.5.2 Universal Surfaces

A *universal surface (US)* is a surface where optimal trajectories join *transversely* from both sides and then have *singular arcs* on it. (See Figure 9.4).

9.5.3 Focal Surfaces

A *focal surface (FS)* is a surface where optimal trajectories join *tangentially* from both sides and then have singular arcs on it. (See Figure 9.5). $\nabla J(\mathbf{x})$ has a jump discontinuity across a focal surface.

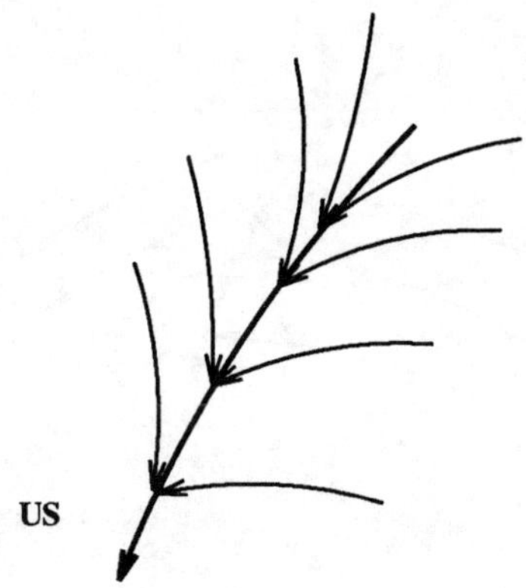

FIGURE 9.4. A universal surface

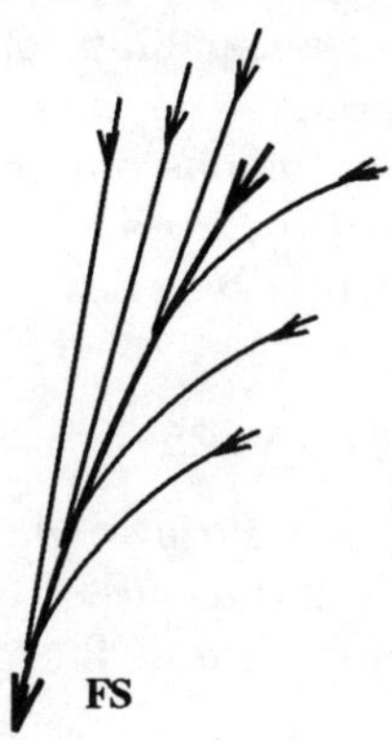

FIGURE 9.5. A focal surface

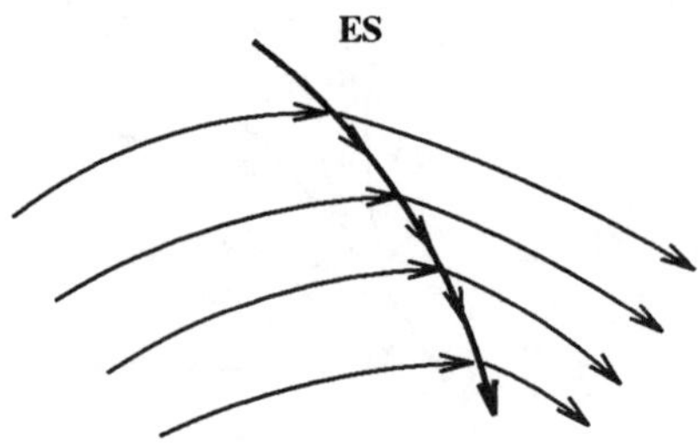

FIGURE 9.6. An equivocal surface.

9.6 Corner Surfaces

9.6.1 General characterization

Corner surfaces are one of the most interesting phenomena in differential games. Corner surfaces are a kind of a hybrid between a switch surface and a dispersal surface. Optimal trajectories arrive from one side and then one player has to make a choice between two alternatives: either to follow the surface as a singular arc or to switch controls very much the same like on a switch surface and cross to the other side as regular trajectories.
This phenomenon represents a capability of a delayable option for one of the players. It resembles a policy, sometimes used in real life conflicts, that is based on the effect of a continuous implicit threat.
The difference between the two types of corner surfaces is again in the nature of the junction of their tributaries.

9.6.2 Equivocal Surfaces

An *equivocal surface (ES)* is a surface where optimal trajectories arrive *transversely* from one side and then either have singular arc on it or leave transversely to the other side (one of the players dominates the choice). (See Figure 9.6).

9.6.3 Switch Envelopes

A *switch envelope (SE)* is a surface where optimal trajectories arrive *tangentially* from one side and then either have a singular arc on it or leaves it *transversely* to the other side. (one of the players dominates the choice). (See Figure 9.7).

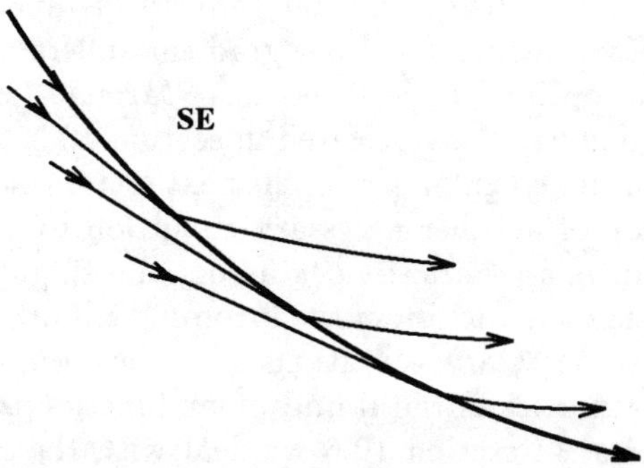

FIGURE 9.7. A switch envelope.

10

Necessary Conditions for Singular Surfaces

10.1 Introduction

In this chapter we treat the *necessary conditions* regarding the properties of solutions to differential games that have discontinuities in the value function $J(\mathbf{x})$ or its gradient $\nabla J(\mathbf{x})$. In section 10.2 we begin with a general property (not limited to differential games!) of functions $J(\mathbf{x})$ at points where they are continuous but their gradient suffers a *jump discontinuity*. The behavior of optimal trajectories near *barriers* (surfaces across which $J(\mathbf{x})$ is discontinuous) is investigated in section 10.3. Section 10.4 presents Isaacs' equations for singular arcs[1]. The next section, section 10.5 is devoted to the derivation of another necessary condition to enable us to compute the value of the *jump discontinuity* across the singular surface. We need this value for the construction of the incoming tributaries that join the singular arc. Those necessary conditions take the form of *junction theorems* that link the properties of the Hamiltonian function to the way tributaries join singular arcs. In section 10.6 we deal with the *adjoint equations for singular arcs*. Those equations are essential for the construction of candidate singular arcs.

Some properties of *regular switch surfaces* are discussed in section 10.7.

We find that along singular arcs one player cannot realize an optimal strategy (within the classical solution concept) unless the information-constraints are violated and he is allowed to base his choice of control on a prior knowledge of his opponent's choice!

In section 10.8 we show how this difficulty is resolved within the extended solution concept. Section 10.10 contains some problems to be worked by the reader.

---

[1] Indeed, the *viscosity form* of Isaacs' equations, (see section 5.5.4), which does not use the notion of *Hamiltonian on a singular arc*, is completely equivalent to those equations. Never the less this formulation fits the derivation of the necessary conditions regarding the junction of tributaries and singular arcs better.

10.2 The Projection Lemma

The *projection lemma* is associated with a general property (not limited only to differential games) of a function $J(\mathbf{x})$ at points $\mathbf{x}$ that belong to a surface where the function is continuous but its gradient, $\nabla J(\mathbf{x})$, has a *jump discontinuity*.

Lemma 10.2.1 (The projection lemma)

Let a smooth (n-1)-dimensional surface D be a locus of points $\mathbf{x}$ at which the function $J(\mathbf{x})$ is continuous but its gradient $\nabla J(\mathbf{x})$ has a jump discontinuity.
Let $\nabla J^{(1)}(\mathbf{x})$ and $\nabla J^{(2)}(\mathbf{x})$ be the limits of $\nabla J(\mathbf{x})$ as it is approached from side-2 and side-1 of the surface D respectively. If $\mathbf{t}(\mathbf{x})$ is any vector tangent to the surface D at the point $\mathbf{x}$ we must have:

$$\nabla J^{(1)}(\mathbf{x}) \cdot \mathbf{t}(\mathbf{x}) = \nabla J^{(2)}(\mathbf{x}) \cdot \mathbf{t}(\mathbf{x}).$$

Proof
We give a heuristic proof.
Let $\boldsymbol{\epsilon}_t \to 0$ be aligned with the vector $\mathbf{t}(\mathbf{x})$.
Adjacent to side-1 we have:

$$J(\mathbf{x} + \boldsymbol{\epsilon}_t) = J(\mathbf{x}) + \nabla J^{(1)}(\mathbf{x}) \cdot \boldsymbol{\epsilon}_t + O[\epsilon_t^2],$$

and adjacent to side-2:

$$J(\mathbf{x} + \boldsymbol{\epsilon}_t) = J(\mathbf{x}) + \nabla J^{(2)}(\mathbf{x}) \cdot \boldsymbol{\epsilon}_t + O[\epsilon_t^2].$$

Comparing the two relations and recalling that they must hold as $\boldsymbol{\epsilon}_t \to 0$ provides the desired result.

•

The *projection lemma* implies that $\nabla J(\mathbf{x})$ has the jump discontinuity *only* along the direction of the normal $\mathbf{n}(\mathbf{x})$ to the surface D at the point $\mathbf{x}$[2].

Corollary 10.2.2

$$\nabla J^{(1)}(\mathbf{x}) = \nabla J^{(2)}(\mathbf{x}) + \beta(\mathbf{x})\mathbf{n}(\mathbf{x}).$$

•

[2]If the dimension of the subspace D is (n-m) then (n-m) independent vectors $\mathbf{t}_i$ span the hyperplane tangent to D at the point $\mathbf{x}$ and m independent vectors span its orthogonal complement. So in general all the projections of $\nabla J(\mathbf{x})$ on a hyperplane tangent to the surfaces D are continuous and the jump discontinuity takes place only in the orthogonal complement of D.

10.3 Open Barriers.

We allow the map of iso-value surfaces in a solution of a differential game to include surfaces across which the value function is discontinuous (*barriers*). Lemma 5.4.3 yields a very important property of optimal trajectories with respect to *barriers*:

Lemma 10.3.1

Optimal trajectories do not cross surfaces across which the function $J(\mathbf{x})$ is discontinuous[3].

Proof
The proof is a direct result of Lemma 5.4.3.

•

From section 5.3 we already know that *closed barriers* that separate the *capture-set* $\mathcal{W}$ from the *escape-set* $\mathcal{F}$ are *semipermeable surfaces*. In the following we shall prove that *barrier* surfaces inside the capture-set (*open barriers*) are also semipermeable surfaces.
Consider a point $\mathbf{x}$ that belong to the smooth part of a barrier B. Let $J^{(1)}(\mathbf{x})$ and $J^{(2)}(\mathbf{x})$ be the limits of $J(\mathbf{x})$ as it is approached from the different sides of the barrier B at the point $\mathbf{x}$.
Let $\mathbf{n}(\mathbf{x})$ be the normal to B at $\mathbf{x}$ pointing to side-2 of the barrier.
With no loss of generality we assume:

$$J^{(1)}(\mathbf{x}) < J^{(2)}(\mathbf{x}).$$

Lemma 10.3.2

If $\mathbf{x}$ is a point on an *open barrier* B and

$$\sigma_B = J^{(2)}(\mathbf{x}) - J^{(1)}(\mathbf{x}) > 0. \tag{10.1}$$

1. A velocity vector $\mathbf{f}(\mathbf{x}, \overset{*}{\mathbf{u}}^{(1)}, \mathbf{v})$ never crosses the *barrier* from side-1 to side-2.

2. A velocity vector $\mathbf{f}(\mathbf{x}, \mathbf{u}, \overset{*}{\mathbf{v}}^{(2)})$ never crosses the *barrier* from side-2 to side-1.

Proof
For the first part of the lemma let us consider a play $[\mathbf{x}^{(1)}, \overset{*}{\mathbf{u}}(\cdot), v(\cdot)]$ where $\mathbf{x}^{(1)}$ is adjacent to $\mathbf{x}$ in side 1 of B.

---

[3]Recall the analogy to a topographic map: we have *cliffs* but *roads* do not cross them.

Let the play be *a play with a slight deviation by player E* as defined in subsection 5.5.1.
Applying the comparison relation (5.9) for this play we get:

$$\int_0^{\Delta t} L[\mathbf{x}(\tau), \overset{*}{\mathbf{u}}(\tau), \mathbf{v}(\tau)]d\tau + J(\mathbf{x}^{(1)} + \Delta\mathbf{x}) \leq J(\mathbf{x}^{(1)}). \tag{10.2}$$

We remember that the function $L(\mathbf{x}, \mathbf{u}, \mathbf{v})$ was assumed to be bounded (See section 2.3),

$$|L(\mathbf{x}, \mathbf{u}, \mathbf{v})| \leq M.$$

It is clear from relation (10.2) that if $\Delta t < \frac{1}{2}\frac{\sigma_B}{M}$ then for $\mathbf{x}^{(1)}$ sufficiently close to $\mathbf{x}$ the state $\mathbf{x}^{(1)} + \Delta\mathbf{x}$ always belongs to side 1 of B.
Recalling that as $\Delta t \to 0$ also $|\Delta\mathbf{x}| \to 0$ and that
$\lim_{\Delta t \to o} \frac{\Delta \mathbf{x}}{\Delta t} = \mathbf{f}(\mathbf{x}^{(1)}, \overset{*}{\mathbf{u}}^{(1)}, \mathbf{v})$
we can get that $\mathbf{f}(\mathbf{x}, \overset{*}{\mathbf{u}}^{(1)}, \mathbf{v})$ never crosses from side 1 to side 2 at $\mathbf{x}$.
The proof that $\mathbf{f}(\mathbf{x}, \mathbf{u}, \overset{*}{\mathbf{v}}^{(1)})$ never crosses from side 2 to side 1 at $\mathbf{x}$ can be worked out similarly.

•

Lemma 10.3.3

Open barriers are semipermeable surfaces.

Proof

It is evident from Lemma 10.3.2 that then $[\overset{*}{\mathbf{u}}^{(1)}(\mathbf{x}), \overset{*}{\mathbf{v}}^{(2)}(\mathbf{x})]$ is a pair of *semipermeable controls* for $\mathbf{x} \in B$[4].

•

Next we shall extend Lemma 10.3.1 in the context of the extended solution concept (defined in section 3.3).

Lemma 10.3.4

Let $[\mathcal{W}, \mathcal{F}, \{\overset{*}{U}_i\}, \{\overset{*}{V}_i\}, J(\mathbf{x})]$ be a solution of a differential game in the extended sense and let B be a *barrier* surface in the capture region $\mathcal{W}$. Given any point $\mathbf{x} \in B$ there exist a number $n(\mathbf{x})$ so that for any $i > n(\mathbf{x})$ *ϵ-optimal trajectories* never cross the surface B at the point $\mathbf{x}$.

Proof

---

[4]The reader should notice that the pair of controls $(\overset{*}{\mathbf{u}}^{(1)}, \overset{*}{\mathbf{v}}^{(2)})$ need not to be an optimal pair in either sides of $\mathbf{x}$.

Let $J^{(1)}(\mathbf{x})$ and $J^{(2)}(\mathbf{x})$ denote the limits of $J(\mathbf{x})$ as $\mathbf{x}$ is approached from side-1 and side-2 of the barrier respectively. Denote:

$$\sigma_B = J^{(2)}(\mathbf{x}) - J^{(1)}(\mathbf{x}) > 0. \tag{10.3}$$

From Lemma 5.4.2 we know that if $\mathbf{x}(t)$ and $\mathbf{x}(s)$ are two points on a ϵ-*optimal trajectory* then given any $\epsilon > 0$ there exists a number $n(\epsilon)$ such that for $i > n(\epsilon)$:

$$J[\mathbf{x}(t)] - J[\mathbf{x}(s)] = \int_t^s L[\mathbf{x}(\tau), \overset{*}{\mathbf{u}}_i(\tau), \overset{*}{\mathbf{v}}_i(\tau)]d\tau + |t-s|O(\epsilon). \tag{10.4}$$

We recall that $L(\mathbf{x}, \mathbf{u}, \mathbf{v})$ was assumed to be bounded (see section 2.6) and that any ϵ-optimal strategy $\overset{*}{\mathbf{u}}_i$ guarantees termination for any trajectory that starts in the capture region $\mathcal{W}$.
If the lemma is not true and for some $\epsilon < \sigma_B$ a ϵ-optimal trajectory crosses at $\mathbf{x} \in B$ from side 1 to side 2 (or from side 2 to side 1) then it is not difficult to show that for $\mathbf{x}(t)$ and $\mathbf{x}(s)$ on different sides of B and for $|t-s|$ sufficiently small relation (10.4) will be in contradiction with relation (10.3).

•

10.4 Isaacs Equations for Singular Arcs

10.4.1 The Hamiltonian on Singular Surfaces

For points $\mathbf{x}$ that belongs to a surface D across which $\nabla J(\mathbf{x})$ suffers a jump discontinuity we have to distinguish between $\nabla J^{(1)}(\mathbf{x})$, the limit of $\nabla J(\mathbf{x})$ as $\mathbf{x}$ is approached from side-1 of D and $\nabla J^{(2)}(\mathbf{x})$, the limit of $\nabla J(\mathbf{x})$ as $\mathbf{x}$ is approached from side-2 of D.
Similarly we shall distinguish between $H(\mathbf{x}, \mathbf{J}_x^{(1)}, \mathbf{u}, \mathbf{v})$, the Hamiltonian at side-1 of $\mathbf{x}$ and $H(\mathbf{x}, \mathbf{J}_x^{(2)}, \mathbf{u}, \mathbf{v})$, the Hamiltonian at side-2 of $\mathbf{x}$ where[5]:

$$H(\mathbf{x}, \mathbf{J}_x^{(1)}, \mathbf{u}, \mathbf{v}) = \nabla J^{(1)}(\mathbf{x}) \cdot \mathbf{f}(\mathbf{x}, \mathbf{u}, \mathbf{v}) + L(\mathbf{x}, \mathbf{u}, \mathbf{v}),$$

$$H(\mathbf{x}, \mathbf{J}_x^{(2)}, \mathbf{u}, \mathbf{v}) = \nabla J^{(2)}(\mathbf{x}) \cdot \mathbf{f}(\mathbf{x}, \mathbf{u}, \mathbf{v}) + L(\mathbf{x}, \mathbf{u}, \mathbf{v}).$$

Lemma 10.4.1

Consider a singular surface D where $J(\mathbf{x})$ is continuous and across which $\nabla J(\mathbf{x})$ has a *jump discontinuity*. Let $\mathbf{n}(\mathbf{x})$be the normal to D at $\mathbf{x}$.
If

$$\mathbf{f}(\mathbf{x}, \mathbf{u}, \mathbf{v}) \cdot \mathbf{n}(\mathbf{x}) = 0,$$

[5] We shall use $\mathbf{J}_x(\mathbf{x})$ as another notation for $\nabla J(\mathbf{x})$ simultaneously for clarity.

Then

$$H[\mathbf{x}, J_x^{(1)}, \mathbf{u}, \mathbf{v}] = H[\mathbf{x}, J_x^{(2)}, \mathbf{u}, \mathbf{v}].$$

Proof

The velocity vector $\mathbf{f}(\mathbf{x}, \mathbf{u}, \mathbf{v})$ is tangent to D at $\mathbf{x}$. By the *projection Lemma* (Lemma 10.2.1) we have:

$$\nabla J^{(2)}(\mathbf{x}) \cdot \mathbf{n}(\mathbf{x}) = \nabla J^{(1)}(\mathbf{x}) \cdot \mathbf{n}(\mathbf{x}),$$

and the lemma follows immediately from the definition of the Hamiltonian.

•

Lemma 10.4.2

If a *singular arc* of an optimal trajectory lies adjacent to side i of a *barrier* or a *singular surface* then:

$$H[\mathbf{x}, J_x^{(i)}, \overset{*}{\mathbf{u}}{}^{(i)}, \overset{*}{\mathbf{v}}{}^{(i)}] = 0.$$

Proof

In Lemma 5.4.4 we showed that $\dot{J}(\mathbf{x})$ exists along an optimal trajectory (including singular arcs!). Evidently:

$$\dot{J}(\mathbf{x}) = \nabla J(\mathbf{x}) \cdot \mathbf{f}(\mathbf{x}, \overset{*}{\mathbf{u}}{}^{(i)}, \overset{*}{\mathbf{v}}{}^{(i)}) = -L(\mathbf{x}, \overset{*}{\mathbf{u}}{}^{(i)}, \overset{*}{\mathbf{v}}{}^{(i)}).$$

•

10.4.2 Isaacs Theorems for Singular Arcs

Relation (5.9) enables us to formulate the equivalent for *Isaacs' equations* for *singular arcs* of optimal trajectories.

Theorem 10.4.3 (Isaacs' theorem for singular arcs)

Let $\mathbf{x}$ be a point on a singular surface D and let $\mathbf{n}(\mathbf{x})$ be the normal to D at $\mathbf{x}$ pointing to the side-2 of the surface D. we have that it is necessary that the optimal controls at $\mathbf{x}$ will satisfy:

$$\begin{aligned}
&H(\mathbf{x}, J_x^{(1)}, \overset{*}{\mathbf{u}}, \mathbf{v}) \leq 0\,, \quad \forall \mathbf{v} \in \mathcal{V} \text{ such that } \mathbf{f}(\mathbf{x}, \overset{*}{\mathbf{u}}, \mathbf{v}) \cdot \mathbf{n}(\mathbf{x}) \leq 0,\\
&H(\mathbf{x}, J_x^{(2)}, \overset{*}{\mathbf{u}}, \mathbf{v}) \leq 0\,, \quad \forall \mathbf{v} \in \mathcal{V} \text{ such that } \mathbf{f}(\mathbf{x}, \overset{*}{\mathbf{u}}, \mathbf{v}) \cdot \mathbf{n}(\mathbf{x}) \geq 0,\\
&H(\mathbf{x}, J_x^{(1)}, \mathbf{u}, \overset{*}{\mathbf{v}}) \geq 0\,, \quad \forall \mathbf{u} \in \mathcal{U} \text{ such that } \mathbf{f}(\mathbf{x}, \mathbf{u}, \overset{*}{\mathbf{v}}) \cdot \mathbf{n}(\mathbf{x}) \leq 0,\\
&H(\mathbf{x}, J_x^{(2)}, \mathbf{u}, \overset{*}{\mathbf{v}}) \geq 0\,, \quad \forall \mathbf{u} \in \mathcal{U} \text{ such that } \mathbf{f}(\mathbf{x}, \mathbf{u}, \overset{*}{\mathbf{v}}) \cdot \mathbf{n}(\mathbf{x}) \geq 0.
\end{aligned} \tag{10.5}$$

Proof

The lemma can be easily derived from relation (5.9).

•

10.4.3 Properties of the Hamiltonian on a Seam between Two Fields Of Regular Optimal Trajectories

Suppose that at both sides of the singular surface D we have fields of regular optimal trajectories. We may consider the surface D as a kind of a *seam surface* between the two regular fields of optimal trajectories that lie at both its sides.

Lemma 10.4.4

$$H[\mathbf{x}, J_x^{(1)}, \overset{*}{\mathbf{u}}^{(1)}, \overset{*}{\mathbf{v}}^{(2)}] \leq 0, \tag{10.6}$$

$$H[\mathbf{x}, J_x^{(2)}, \overset{*}{\mathbf{u}}^{(1)}, \overset{*}{\mathbf{v}}^{(2)}] \geq 0, \tag{10.7}$$

$$H[\mathbf{x}, J_x^{(1)}, \overset{*}{\mathbf{u}}^{(2)}, \overset{*}{\mathbf{v}}^{(1)}] \geq 0, \tag{10.8}$$

$$H[\mathbf{x}, J_x^{(2)}, \overset{*}{\mathbf{u}}^{(2)}, \overset{*}{\mathbf{v}}^{(1)}] \leq 0. \tag{10.9}$$

Proof

Let us prove relation (10.6) (the proofs of relation (10.7), relation (10.8), and relation (10.9) are almost similar and will be omitted).

At all points $\mathbf{x}^{(1)}$ of field-1 of the regular optimal trajectories adjacent to the seam surface D it is necessary (by Isaacs' theorems with respect to field-1) that:

$$H[\mathbf{x}^{(1)}, J_x, \overset{*}{\mathbf{u}}^{(1)}, \mathbf{v}] \leq 0\ , \quad \text{for all } \mathbf{v} \in \mathcal{V}. \tag{10.10}$$

As we approach a point $\mathbf{x}$ on the seam surface D from side-1 we must therefore have:

$$\lim_{\mathbf{x}^{(1)} \to \mathbf{x}} H[\mathbf{x}^{(1)}, J_x, \overset{*}{\mathbf{u}}^{(1)}, \mathbf{v}] = H[\mathbf{x}, J_x^{(1)}, \overset{*}{\mathbf{u}}^{(1)}, \mathbf{v}] \leq 0, \quad \text{for all } \mathbf{v} \in \mathcal{V}. \tag{10.11}$$

•

The lemma may be interpreted as follows:

relation (10.6) means that $\overset{*}{\mathbf{v}}^{(2)}$ may be considered as a *deviation* from the optimal controls by player E with respect to field-1.

relation (10.7) means that $\overset{*}{\mathbf{u}}^{(1)}$ may be considered as a *deviation* from the optimal controls by player P with respect to field-2.

relation (10.8) means that $\overset{*}{\mathbf{u}}^{(2)}$ may be considered as a *deviation* from the optimal controls by player P with respect to field-1.

relation (10.9) means that $\overset{*}{\mathbf{v}}^{(1)}$ may be considered as a *deviation* from the optimal controls by player E with respect to field-2.

•

From Lemma 10.4.4 we get the following important result:

Lemma 10.4.5

At a point $\mathbf{x}$ on a seam surface D

$$H[\mathbf{x}, J_x^{(1)}, \overset{*}{\mathbf{u}}^{(1)}, \overset{*}{\mathbf{v}}^{(2)}] - H[\mathbf{x}, J_x^{(2)}, \overset{*}{\mathbf{u}}^{(1)}, \overset{*}{\mathbf{v}}^{(2)}] = \beta(\mathbf{x})\mathbf{n}(\mathbf{x}) \cdot \mathbf{f}(\mathbf{x}, \overset{*}{\mathbf{u}}^{(1)}, \overset{*}{\mathbf{v}}^{(2)}) \leq 0,$$

$$H[\mathbf{x}, J_x^{(1)}, \overset{*}{\mathbf{u}}^{(2)}, \overset{*}{\mathbf{v}}^{(1)}] - H[\mathbf{x}, J_x^{(2)}, \overset{*}{\mathbf{u}}^{(2)}, \overset{*}{\mathbf{v}}^{(1)}] = \beta(\mathbf{x})\mathbf{n}(\mathbf{x}) \cdot \mathbf{f}(\mathbf{x}, \overset{*}{\mathbf{u}}^{(2)}, \overset{*}{\mathbf{v}}^{(1)}) \geq 0.$$

Or equivalently:

if $\beta(\mathbf{x}) \neq 0$ the velocity vectors $\mathbf{f}(\mathbf{x}, \overset{*}{\mathbf{u}}^{(1)}, \overset{*}{\mathbf{v}}^{(2)})$ and $\mathbf{f}(\mathbf{x}, \overset{*}{\mathbf{u}}^{(2)}, \overset{*}{\mathbf{v}}^{(1)})$ do not lead to the same side of the surface D.

Proof

The proof is a direct result of the *projection lemma* (Lemma 10.2.1):

$$\nabla J^{(1)}(\mathbf{x}) = \nabla J^{(2)}(\mathbf{x}) + \beta(\mathbf{x})\mathbf{n}(\mathbf{x}), \tag{10.12}$$

and of Lemma 10.4.4.

•

10.5 Necessary Conditions at Junctions Between Regular and Singular Segments of Optimal Trajectories

10.5.1 Controls Along Singular Arcs

At a *singular arc* the optimal trajectory is confined to lie along a certain surface Q in the game-set.
For example:

1. The singular arc is confined to lie on the NUP. (This is sometimes called a *safe contact* arc).
2. The singular arc is confined to be adjacent to a certain side of a *barrier surface* B. (This is also a case of a *safe contact* arc).
3. The singular arc is confined to lie on a *seam surface* D that separates two different fields of regular optimal trajectories.

There is a phenomenon that is common to all three cases:

> One player (for example player P) dominates in the sense that the other player (for example player E) can realize his optimal controls along the surface Q *only* by applying a *nonadmissible strategy* that involves the knowledge of the current choice of control by his opponent.

A typical situation is characterized by:

1.
$$\mathbf{f}[\mathbf{x}, \overset{*}{\mathbf{u}}^{(i)}, \overset{*}{\mathbf{v}}^{(i)}] \cdot \mathbf{n}(\mathbf{x}) \leq 0, \tag{10.13}$$

   where $\mathbf{n}(\mathbf{x})$ is the normal to the surface Q pointing into side-i of Q at $\mathbf{x}$.
   Relation (10.13) imply that regular optimal trajectories of field-i head *toward* the surface Q.

2.
$$\mathbf{f}[\mathbf{x}, \mathbf{u}, \overset{*}{\mathbf{v}}(\mathbf{u})] \cdot \mathbf{n}(\mathbf{x}) = 0 \ , \quad \text{for all} \quad \mathbf{u} \in \mathcal{U}_1, \tag{10.14}$$

   where the set of controls $\mathcal{U}_1$ is a subset[6] of the set of admissible controls $\mathcal{U}$ and $\overset{*}{\mathbf{u}}^{(i)} \in \mathcal{U}_1$ so that:

$$\mathbf{f}[\mathbf{x}, \overset{*}{\mathbf{u}}^{(i)}, \mathbf{v}(\overset{*}{\mathbf{u}}^{(i)})] \cdot \mathbf{n}(\mathbf{x}) = 0. \tag{10.15}$$

   Relation (10.14) means that if player E is assumed to know the current choice of control by player P then for a certain set of choices by player P he can force the trajectory to stay on the singular surface by control laws of the type $\mathbf{v} = \mathbf{v}(\mathbf{u})$.

We shall denote the *optimal* control laws of player E and player P at a singular surface as $\tilde{\mathbf{v}} = \tilde{\mathbf{v}}(\mathbf{u})$ and $\tilde{\mathbf{u}}$ respectively. The pair $[\tilde{\mathbf{u}}, \tilde{\mathbf{v}}(\tilde{\mathbf{u}})]$ corre-sponds to the case when both players realize their optimal control laws and the resulting trajectory along the singular surface is a *singular arc* segment of the optimal trajectory[7].

In general we have to assume that at $\mathbf{x}$ the pair $(\overset{*}{\mathbf{u}}^{(i)}, \overset{*}{\mathbf{v}}^{(i)})$ of optimal controls with respect to the *regular* field of optimal trajectories at side-i of $\mathbf{x}$ of the surface Q is not the same as the pair $[\tilde{\mathbf{u}}, \tilde{\mathbf{v}}(\tilde{\mathbf{u}})]$ of optimal controls at $\mathbf{x}$ with respect to the *singular arc* that lies on the surface Q.

Lemma 10.5.1

---

[6]The set $\mathcal{U}_1$ depends on the properties of the specific game and on the type of the singular surface.

[7]The control laws $\mathbf{v} = \mathbf{v}(\mathbf{u})$ are clearly nonadmissible strategies (see subsection 2.7.5) and their introduction deserves an explanation. As a matter of fact we shall later argue that the singular arc and the corresponding nonadmissible control laws are asymptotic *limit* of ϵ- optimal trajectories (as $\epsilon \to 0$) realized by admissible control laws in the sense of the extended solution concept where player E can guarantee himself an arbitrarily small proximity to the value that corresponds to the use of the nonadmissible strategies.

For points $\mathbf{x}$ on the singular arc we have:

$$H[\mathbf{x}, \mathbf{J}^{(\mathbf{i})}_{\mathbf{x}}, \tilde{\mathbf{u}}, \tilde{\mathbf{v}}(\tilde{\mathbf{u}})] = \mathbf{0} \tag{10.16}$$

Proof
The proof is an immediate result of the fact that both Lemma 5.4.3 and Lemma 5.4.4 are valid on both the regular parts and the singular arcs of optimal trajectories.

•

Lemma 10.5.2

$\tilde{\mathbf{u}}$ solves the problem:

$$\begin{aligned} &\min H[\mathbf{x}, \mathbf{J}^{(\mathbf{i})}_{\mathbf{x}}(\mathbf{x}), \mathbf{u}, \tilde{\mathbf{v}}(\mathbf{u})], \\ &\mathbf{u} \in \{\, \mathbf{u} \mid \mathbf{u} \in \mathcal{U}_1 \ , \ \mathbf{f}[\mathbf{x}, \mathbf{u}, \tilde{\mathbf{v}}(\mathbf{u})] \cdot \mathbf{n}(\mathbf{x}) = 0\}. \end{aligned} \tag{10.17}$$

Proof
Relation (5.9) is valid also for deviations that are confined to the surface Q. From relation (5.9) we have that for sufficiently small Δt the following will be true for all $\mathbf{u} \in \mathcal{U}_1$:

$$J(\mathbf{x} + \mathbf{\Delta x}) + \int_0^{\Delta t} L\{\mathbf{x}(\tau), \mathbf{u}(\tau), \tilde{\mathbf{v}}[\mathbf{u}(\tau)]\} d\tau \geq J(\mathbf{x}). \tag{10.18}$$

Let us recall that:

1. Both $\mathbf{x}$ and $\mathbf{x} + \Delta\mathbf{x}$ lie on the surface Q.
2. $L(\mathbf{x}, \mathbf{u}, \mathbf{v})$ was assumed to be bounded. (See section 2.3).
3. As $\Delta t \to 0$, also $|\mathbf{\Delta x}| \to \mathbf{0}$.
4. As $\Delta t \to 0$,

   $$\lim_{\Delta t \to 0} \frac{\Delta x}{\Delta t} = \mathbf{f}[\mathbf{x}, \mathbf{u}, \tilde{\mathbf{v}}(\mathbf{u})],$$

   and that for $\mathbf{u} \in \mathcal{U}_1$:

   $$\mathbf{f}[\mathbf{x}, \mathbf{u}, \tilde{\mathbf{v}}(\mathbf{u})] \cdot \mathbf{n}(\mathbf{x}) = 0.$$

Evidently relation (10.18) can now be rewritten in a differential form:

$$\begin{aligned} &\nabla J^{(i)}(\mathbf{x}) \cdot \mathbf{f}[\mathbf{x}, \mathbf{u}, \tilde{\mathbf{v}}(\mathbf{u})] \geq 0, \\ &\forall \mathbf{u} \in \{\, \mathbf{u} \mid \mathbf{u} \in \mathcal{U}_1 \ , \ \mathbf{f}[\mathbf{x}, \mathbf{u}, \tilde{\mathbf{v}}(\mathbf{u})] \cdot \mathbf{n}(\mathbf{x}) = 0\}. \end{aligned} \tag{10.19}$$

Which is the same as

$$H[\mathbf{x}, \mathbf{J}^{(\mathbf{i})}_{\mathbf{x}}(\mathbf{x}), \mathbf{u}, \tilde{\mathbf{v}}(\mathbf{u})] \geq \mathbf{0}, \qquad \forall \mathbf{u} \in \mathcal{U}_{\mathbf{1}}. \tag{10.20}$$

•

Lemma 10.5.3

$\tilde{\mathbf{v}}(\tilde{\mathbf{u}})$ solves the problem:

$$\max H[\mathbf{x}, \mathbf{J}_{\mathbf{x}}^{(\mathbf{i})}(\mathbf{x}), \tilde{\mathbf{u}}, \mathbf{v}(\tilde{\mathbf{u}})],$$
$$\mathbf{v} \in \{\, \mathbf{v} \mid \mathbf{v} \in \mathcal{V} \,,\ \mathbf{f}[\mathbf{x}, \tilde{\mathbf{u}}, \mathbf{v}(\tilde{\mathbf{u}})] \cdot \mathbf{n}(\mathbf{x}) = 0\}. \tag{10.21}$$

Proof
The proof resembles that of Lemma 10.5.2 and its details are left to the reader.

•

Lemma 10.5.4

At the point $\mathbf{x}$ where an optimal trajectory that *arrives* from side-i of the surface Q joins a *singular arc* that lies on Q we have:

$$H[\mathbf{x}, \mathbf{J}_{\mathbf{x}}^{(\mathbf{i})}, \overset{*}{\mathbf{u}}{}^{(\mathbf{i})}, \overset{*}{\mathbf{v}}{}^{(\mathbf{i})}] = \mathbf{H}[\mathbf{x}, \mathbf{J}_{\mathbf{x}}^{(\mathbf{i})}, \overset{*}{\mathbf{u}}{}^{(\mathbf{i})}, \tilde{\mathbf{v}}(\overset{*}{\mathbf{u}}{}^{(\mathbf{i})})] = \mathbf{H}[\mathbf{x}, \mathbf{J}_{\mathbf{x}}^{(\mathbf{i})}, \tilde{\mathbf{u}}, \tilde{\mathbf{v}}(\tilde{\mathbf{u}})] = \mathbf{0}. \tag{10.22}$$

Proof
At any point $\mathbf{x}$ we may consider the pair of controls:

$$[\overset{*}{\mathbf{u}}{}^{(i)}, \tilde{\mathbf{v}}(\overset{*}{\mathbf{u}}{}^{(i)})],$$

as a *deviation*, by player P, with respect to the minimization problem of Lemma 10.5.2:

$$H[\mathbf{x}, \mathbf{J}_{\mathbf{x}}^{(\mathbf{i})}, \overset{*}{\mathbf{u}}{}^{(\mathbf{i})}, \tilde{\mathbf{v}}(\overset{*}{\mathbf{u}}{}^{(\mathbf{i})})] \geq \mathbf{H}[\mathbf{x}, \mathbf{J}_{\mathbf{x}}^{(\mathbf{i})}, \tilde{\mathbf{u}}, \tilde{\mathbf{v}}(\tilde{\mathbf{u}})] = \mathbf{0}, \tag{10.23}$$

and at the same time we may consider this pair as a *deviation* , by player E, with respect to the optimal controls of the regular field-i:

$$H[\mathbf{x}, \mathbf{J}_{\mathbf{x}}^{(\mathbf{i})}, \overset{*}{\mathbf{u}}{}^{(\mathbf{i})}, \tilde{\mathbf{v}}(\overset{*}{\mathbf{u}}{}^{(\mathbf{i})})] \leq \mathbf{H}[\mathbf{x}, \mathbf{J}_{\mathbf{x}}^{(\mathbf{i})}, \overset{*}{\mathbf{u}}{}^{(\mathbf{i})}, \overset{*}{\mathbf{v}}{}^{(\mathbf{i})}] = \mathbf{0}. \tag{10.24}$$

$H[\mathbf{x}, \mathbf{J}_{\mathbf{x}}^{(\mathbf{i})}, \overset{*}{\mathbf{u}}{}^{(\mathbf{i})}, \tilde{\mathbf{v}}(\overset{*}{\mathbf{u}}{}^{(\mathbf{i})})]$ *squeezes* the inequalities (10.23) and (10.24) to be *equalities*! and the lemma follows.

•

Lemma 10.5.4 has two straight forward corollaries:

Corollary 10.5.5

Both $\overset{*}{\mathbf{u}}{}^{(i)}$ and $\tilde{\mathbf{u}}$ solve the optimization problem:

$$\min H[\mathbf{x}, \mathbf{J}_{\mathbf{x}}^{(\mathbf{i})}, \mathbf{u}, \tilde{\mathbf{v}}(\mathbf{u})],$$
$$\mathbf{u} \in \{\, \mathbf{u} \mid \mathbf{u} \in \mathcal{U}_1 \,,\ \mathbf{f}[\mathbf{x}, \mathbf{u}, \tilde{\mathbf{v}}(\mathbf{u})] \cdot \mathbf{n}(\mathbf{x}) = 0\}. \tag{10.25}$$

•

Corollary 10.5.6

Both $\overset{*}{\mathbf{v}}^{(i)}$ and $\tilde{\mathbf{v}}(\overset{*}{\mathbf{u}}^{(i)})$ solve the optimization problem:

$$\max_{\mathbf{v} \in \mathcal{V}} H[\mathbf{x}, \mathbf{J}_{\mathbf{x}}^{(\mathbf{i})}, \overset{*}{\mathbf{u}}^{(\mathbf{i})}, \mathbf{v}]. \tag{10.26}$$

•

10.5.2 The Junction Conditions

When we discussed the *hodograph representation* in section 5.5.2 we saw that the number of solutions of the optimization problems in Corollaries 10.5.6 and 10.5.5 depend on the nature of the Hamiltonian functional. By using similar arguments together with Lemma 10.5.4 we can deduce the following lemmas concerning *necessary conditions* on the nature of the junction of regular optimal trajectories arriving at a singular surface from side-i to singular arcs that adhere to the singular surface for a while.

Lemma 10.5.7 (Junction condition 1)

If the minimization problem:

$$\begin{gathered} \min H[\mathbf{x}, \mathbf{J}_{\mathbf{x}}^{(\mathbf{i})}, \mathbf{u}, \tilde{\mathbf{v}}(\mathbf{u})], \\ \mathbf{u} \in \{\, \mathbf{u} \mid \mathbf{u} \in \mathcal{U}_1 \;,\; \mathbf{f}[\mathbf{x}, \mathbf{u}, \tilde{\mathbf{v}}(\mathbf{u})] \cdot \mathbf{n}(\mathbf{x}) = 0\}, \end{gathered} \tag{10.27}$$

has a unique solution (at side-i) then:

$$\overset{*}{\mathbf{u}}^{(i)} = \tilde{\mathbf{u}}. \tag{10.28}$$

Proof
The proof is a direct result of Lemma 10.5.4.

•

Lemma 10.5.8 (Junction condition 2)

If the maximization problem:

$$\max_{\mathbf{v} \in \mathcal{V}} H[\mathbf{x}, \mathbf{J}_{\mathbf{x}}^{(\mathbf{i})}, \overset{*}{\mathbf{u}}^{(\mathbf{i})}, \mathbf{v}], \tag{10.29}$$

has a unique solution then:

$$\mathbf{f}(\mathbf{x}, \overset{*}{\mathbf{u}}^{(i)}, \overset{*}{\mathbf{v}}^{(i)}) \cdot \mathbf{n}(\mathbf{x}) = 0. \tag{10.30}$$

(The regular optimal trajectories of field-i that arrive and join the the singular arc at $\mathbf{x}$ are tangent to the singular surface Q[8]).

Proof
By relation 10.15 we have:

$$\mathbf{f}[\mathbf{x}, \overset{*}{\mathbf{u}}^{(i)}, \mathbf{v}(\overset{*}{\mathbf{u}}^{(i)})] \cdot \mathbf{n}(\mathbf{x}) = 0. \tag{10.31}$$

We also have, by Corollary 10.5.6, that:

$$\overset{*}{\mathbf{v}}^{(i)} = \tilde{\mathbf{v}}(\overset{*}{\mathbf{u}}^{(i)}) \tag{10.32}$$

Substituting from equation (10.32) to equation (10.31) we obtain

$$\mathbf{f}(\mathbf{x}, \overset{*}{\mathbf{u}}^{(i)}, \overset{*}{\mathbf{v}}^{(i)}) \cdot \mathbf{n}(\mathbf{x}) = 0. \tag{10.33}$$

•

Lemma 10.5.9 (Junction condition 3)

If both the problem:

$$\begin{aligned} &\min H[\mathbf{x}, \mathbf{J}_{\mathbf{x}}^{(\mathbf{i})}(\mathbf{x}), \mathbf{u}, \tilde{\mathbf{v}}(\mathbf{u})], \\ &\mathbf{u} \in \{\ \mathbf{u} \mid \mathbf{u} \in \mathcal{U}_1\ ,\ \mathbf{f}[\mathbf{x}, \mathbf{u}, \tilde{\mathbf{v}}(\mathbf{u})] \cdot \mathbf{n}(\mathbf{x}) = 0\}, \end{aligned} \tag{10.34}$$

and the problem:

$$\begin{aligned} &\max H[\mathbf{x}, \mathbf{J}_{\mathbf{x}}^{(\mathbf{i})}(\mathbf{x}), \overset{*}{\mathbf{u}}^{(\mathbf{i})}, \mathbf{v}], \\ &\mathbf{v} \in \mathcal{V} \end{aligned} \tag{10.35}$$

have unique solutions then:

$$[\overset{*}{\mathbf{u}}^{(i)}, \overset{*}{\mathbf{v}}^{(i)}] = [\tilde{\mathbf{u}}, \tilde{\mathbf{v}}(\tilde{\mathbf{u}})]. \tag{10.36}$$

(The regular optimal trajectory and the singular arc join *smoothly*).

Proof
We have:

$$\begin{aligned} &\overset{*}{\mathbf{u}}^{(i)} = \tilde{\mathbf{u}}, \\ &\overset{*}{\mathbf{v}}^{(i)} = \tilde{\mathbf{v}}(\overset{*}{\mathbf{u}}^{(i)}) = \tilde{\mathbf{v}}(\tilde{\mathbf{u}}), \end{aligned} \tag{10.37}$$

and the lemma follows.

•

[8] The reader should notice that if the dimension of the singular surface is $n \geq 2$ the junction between the arriving regular optimal trajectory and the singular arc need not be smooth.

Lemma 10.5.10 (Junction condition 4.)

If

$$\mathbf{f}[\mathbf{x}, \overset{*}{\mathbf{u}}^{(i)}, \overset{*}{\mathbf{v}}^{(i)}] \cdot \mathbf{n}(\mathbf{x}) < 0, \tag{10.38}$$

(Regular optimal trajectories of field-i arrive at the surface Q *transversely*) then $H[\mathbf{x}, \mathbf{J}_{\mathbf{x}}^{(\mathbf{i})}, \overset{*}{\mathbf{u}}^{(\mathbf{i})}, \mathbf{v}]$ is necessarily linear in component(s) of the control $\mathbf{v}$ and the switch function $A[\mathbf{x}, \mathbf{J}_{\mathbf{x}}^{(\mathbf{i})}, \overset{*}{\mathbf{u}}^{(\mathbf{i})}, \mathbf{v}]$ of those component(s) vanishes at the junction point $\mathbf{x}$.

Proof

See Corollary 5.5.5

•

10.6 Adjoint Equations for Singular Arcs

Optimal trajectories may have *singular arcs* that adhere to singular surfaces across which $\nabla J(\mathbf{x})$ suffers a jump-discontinuity.

At points $\mathbf{x}$ on singular arcs that lie in a (n-1)-dimensional *singular surface* D we have that $\mathbf{f}(\mathbf{x}, \overset{*}{\mathbf{u}}, \overset{*}{\mathbf{v}})$ is tangent to D and that $J_{xx}(\mathbf{x})$ does not exist ($J(\mathbf{x})$ is not C^2). We cannot carry the arguments that led to the *adjoint equations*, (n relations between the n components of $\mathbf{J}_x$), the way it was done in section 5.6.

In the following we shall show (following *P. Bernhard* in [Ber77]) that the restriction of the value function $J(\mathbf{x})$ to the singular surface yields a C^2 (n-1)-dimensional function $W(\mathbf{m})$, which can lead, by arguments similar to those of section 5.6, to adjoint equations ((n-1) independent relations) that hold on singular arcs.

Consider a (n-1)-dimensional surface $D(\mathbf{x}) = 0$,

Let us parametrize the surface D by (n-1) parameters:

$$\mathbf{m} = [l_1, \ldots, l_{n-2}, \tau].$$

We designate

$$\mathbf{l} = [l_1, \ldots, l_{n-2}],$$

so

$$\mathbf{x} = \mathbf{y}(\mathbf{m}) = \mathbf{y}(\mathbf{l}, \tau).$$

Let the set of equations $\mathbf{h}(\mathbf{l})$ be defined as:

$$h_i(\mathbf{l}) = y_i(\mathbf{l}, 0)$$

$\mathbf{h}(\mathbf{l})$ describes a (n-2)-dimensional curve S that lies on the surface D.

Let us consider the scalar functions $W(\mathbf{m})$ and $T(\mathbf{l})$ which are the restriction of the value function $J(\mathbf{x})$ to the surface D and to the curve S

respectively,

$$W(\mathbf{m}) = J[\mathbf{y}(\mathbf{m})] \ , \quad \text{for } \mathbf{x} \in D,$$

$$T(\mathbf{l}) = J[\mathbf{h}(\mathbf{l})] \ , \qquad \text{for } \mathbf{x} \in S.$$

In the following we shall assume that the function $W(\mathbf{m})$ is C^2 on D though the function $J(\mathbf{x})$ is not C^2 there.
Consider a singular arc of an optimal trajectory that starts at $t = 0$ at a point $\mathbf{x}(0) \in S$ and stays on the singular surface D for a while.
At time t:

$$\mathbf{x}(t) = \mathbf{y}[\mathbf{m}(t)],$$

$$J[\mathbf{x}(t)] = W[\mathbf{m}(t)].$$

Let us now adopt the so called *summation convention*. If an index appears twice in any factor it will mean that the factor is summed over the range of that index. For example: $a_i b_i$ will mean $\sum_i a_i b_i$.
The optimal velocity vector $\overset{*}{\mathbf{f}}$ will be related to the trajectory through the relations

$$\overset{*}{\mathbf{f}}_i \; [y(m)] = [\frac{\partial y_l}{\partial m_i}][\frac{dm_i}{dt}], \tag{10.39}$$

and the function W relates to the function J by

$$\frac{\partial W(\mathbf{m})}{\partial m_i} = [\frac{\partial J^{(k)}}{\partial y_l}][\frac{\partial y_l}{\partial m_i}] \ , \quad \text{for} \quad [k = 1, 2], \tag{10.40}$$

where at $\mathbf{x} \in D$ we define $[\frac{\partial J^{(k)}}{\partial y_l}]$ as the limit of $[\frac{\partial J}{\partial y_l}]$ as $\mathbf{x}$ is approached from side-k ($k = 1, 2$) of the singular surface D.
Notice that relation (10.40) is true for both $k = 1$ and $k = 2$ but this is no wonder because relation (10.40) is equivalent to Lemma 10.2.1 (the *projection Lemma*).

Lemma 10.6.1

If $\frac{\partial^2 W}{\partial m_i \partial m_j}$ exists, (or equivalently if $W(\mathbf{m})$ is a C^2 function), then:

$$[\frac{\partial^2 J^{(k)}}{\partial y_l \partial y_n}][\frac{\partial y_n}{\partial m_j}][\frac{\partial y_l}{\partial m_i}] = [\frac{\partial^2 J^{(k)}}{\partial y_l \partial y_n}][\frac{\partial y_n}{\partial m_i}][\frac{\partial y_l}{\partial m_j}].$$

Proof
Differentiating relation (10.40) with respect to m_j we get:

$$\frac{\partial^2 W}{\partial m_i \partial m_j} = [\frac{\partial^2 J^{(k)}}{\partial y_l \partial y_n}][\frac{\partial y_n}{\partial m_j}][\frac{\partial y_l}{\partial m_i}] + [\frac{\partial J^{(k)}}{\partial y_l}][\frac{\partial^2 y_l}{\partial m_i \partial m_j}]. \tag{10.41}$$

Let us now interchange i with j in equation (10.41) and equation (10.40). We get:

$$\frac{\partial^2 W}{\partial m_j \partial m_i} = [\frac{\partial^2 J^{(k)}}{\partial y_l \partial y_n}][\frac{\partial y_n}{\partial m_i}][\frac{\partial y_l}{\partial m_j}] + [\frac{\partial J^{(k)}}{\partial y_l}][\frac{\partial^2 y_l}{\partial m_j \partial m_i}]. \tag{10.42}$$

Comparing relation (10.41) with relation (10.42), recalling that the consecutive differentiation with respect to m_i and m_j commute, our assertion follows.

•

Lemma 10.6.2 (Adjoint equations for singular arcs)

Let D be a singular surface parametrized by $\mathbf{m} = [l_1, \ldots, l_{n-2}, \tau]$ and let $W(\mathbf{m})$ be the restriction of $J(\mathbf{x})$ to the surface D.
If $\frac{\partial^2 W}{\partial m_j \partial m_i}$ exists then along a singular arc of an optimal trajectory that lies on D we have[9]:

$$[\frac{\partial y_l}{\partial m_j}][\frac{\partial H^{(k)}}{\partial y_l} + \frac{d}{dt}(\frac{\partial J^{(k)}}{\partial y_l})] = 0, \tag{10.43}$$

which is also equivalent to:

$$\frac{d}{dt}\nabla J^{(k)}(\mathbf{y}) = -\frac{\partial H(\mathbf{y}, J_y^{(k)}, \overset{*}{\mathbf{u}}, \overset{*}{\mathbf{v}})}{\partial \mathbf{y}} + \Gamma(t)\mathbf{n}(\mathbf{y}), \tag{10.44}$$

where $\mathbf{n}(\mathbf{y})$ is the normal to D at $\mathbf{x} = \mathbf{y}(\mathbf{m})$.

Proof
Let $\mathbf{y}(\mathbf{m}) \in D$ be a point on a *singular arc* that lie on the singular surface D.
By Lemma 10.4.2:

$$H\{\mathbf{y}(\mathbf{m}), J_y^{(k)}[\mathbf{y}(\mathbf{m})], \overset{*}{\mathbf{u}}[\mathbf{y}(\mathbf{m})], \overset{*}{\mathbf{v}}[\mathbf{y}(\mathbf{m})]\} = 0. \tag{10.45}$$

Let $\mathbf{y}(\mathbf{m}+\Delta\mathbf{m}) \in D$ be a neighboring point. We assume that it also belongs to some *singular arc* that lie on the surface D hence also:

$$H\{\mathbf{y}(\mathbf{m}+\Delta\mathbf{m}), J_y^{(k)}[\mathbf{y}(\mathbf{m}+\Delta\mathbf{m})], \overset{*}{\mathbf{u}}[\mathbf{y}(\mathbf{m}+\Delta\mathbf{m})], \overset{*}{\mathbf{v}}[\mathbf{y}(\mathbf{m}+\Delta\mathbf{m})]\} = 0. \tag{10.46}$$

A necessary condition that relation (10.46) be valid for arbitrary small $\Delta\mathbf{m}$ is:

$$\begin{aligned} 0 \;=\; & [\frac{\partial H^{(k)}}{\partial y_l}][\frac{\partial y_l}{\partial m_j}] + [\overset{*}{f_l}][\frac{\partial^2 J^{(k)}}{\partial y_l \partial y_n}][\frac{\partial y_n}{\partial m_j}] + \\ & +[\frac{\partial H^{(k)}}{\partial \overset{*}{\mathbf{u}}_p}][\frac{\partial \overset{*}{\mathbf{u}}_p}{\partial y_n}][\frac{\partial y_n}{\partial m_j}] + [\frac{\partial H^{(k)}}{\partial \overset{*}{\mathbf{v}}_q}][\frac{\partial \overset{*}{\mathbf{v}}_q}{\partial y_n}][\frac{\partial y_n}{\partial m_j}]. \end{aligned} \tag{10.47}$$

[9]We use the notation $H^{(k)}$ as a substitute for

$$H\{\mathbf{y}(\mathbf{m}), J_y^{(k)}[\mathbf{y}(\mathbf{m})], \overset{*}{\mathbf{u}}[\mathbf{y}(\mathbf{m})], \overset{*}{\mathbf{v}}[\mathbf{y}(\mathbf{m})]\}.$$

The reader should notice that the displacement $\Delta\mathbf{m}$ is spanned by (n-1) base-vectors that lie on a plane tangent to the surface D at $\mathbf{y}(\mathbf{m})$. For this reason equation (10.47) represents only (n-1) independent relations (compared to n independent relations that we got in the *regular* case of relation (5.32) where $\Delta\mathbf{x}$ was allowed to have an arbitrary direction).
We can modify the arguments of Theorem 5.6.1 to show that[10]

$$[\frac{\partial H^{(k)}}{\partial \overset{*}{\mathbf{u}}_p}][\frac{\partial \overset{*}{\mathbf{u}}_p}{\partial y_n}][\frac{\partial y_n}{\partial m_j}] = 0, \tag{10.48}$$

$$[\frac{\partial H^{(k)}}{\partial \overset{*}{\mathbf{v}}_q}][\frac{\partial \overset{*}{\mathbf{v}}_q}{\partial y_n}][\frac{\partial y_n}{\partial m_j}] = 0. \tag{10.49}$$

So the differential equivalent of relation (10.46) is

$$[\frac{\partial H^{(k)}}{\partial y_l}][\frac{\partial y_l}{\partial m_j}] + [\overset{*}{f}_l][\frac{\partial^2 J^{(k)}}{\partial y_l \partial y_n}][\frac{\partial y_n}{\partial m_j}] = 0. \tag{10.50}$$

Substituting the relation (10.39) for $\overset{*}{f}_i\ [y(m)]$ in equation (10.50) we get:

$$[\frac{\partial H^{(k)}}{\partial y_l}][\frac{\partial y_l}{\partial m_j}] + [\frac{\partial y_l}{\partial m_i}][\frac{dm_i}{dt}][\frac{\partial^2 J^{(k)}}{\partial y_l \partial y_n}][\frac{\partial y_n}{\partial m_j}] = 0. \tag{10.51}$$

Recalling Lemma 10.6.1 we can rewrite relation (10.51) as:

$$[\frac{\partial H^{(k)}}{\partial y_l}][\frac{\partial y_l}{\partial m_j}] + [\frac{\partial y_l}{\partial m_j}][\frac{dm_i}{dt}][\frac{\partial^2 J^{(k)}}{\partial y_l \partial y_n}][\frac{\partial y_n}{\partial m_i}] = 0, \tag{10.52}$$

or:

$$\left[[\frac{\partial H^{(k)}}{\partial y_l}] + [\frac{dm_i}{dt}][\frac{\partial^2 J^{(k)}}{\partial y_l \partial y_n}][\frac{\partial y_n}{\partial m_i}]\right][\frac{\partial y_l}{\partial m_j}] = 0, \tag{10.53}$$

which is the same as:

$$\left[[\frac{\partial H^{(k)}}{\partial y_l}] + \frac{d}{dt}[\frac{\partial J^{(k)}}{\partial y_l}]\right][\frac{\partial y_l}{\partial m_j}] = 0. \tag{10.54}$$

Notice that $[\frac{\partial y_l}{\partial m_j}]$ is a set of (n-1) vectors that span a hyperplane that is tangent to D at $\mathbf{y}(\mathbf{m})$ so that relation (10.54) is also equivalent to:

$$\frac{d}{dt}\nabla J^{(k)}(\mathbf{y}) = -\frac{\partial H(\mathbf{y}, J_y^{(k)}, \overset{*}{\mathbf{u}}, \overset{*}{\mathbf{v}})}{\partial \mathbf{y}} + \Gamma(t)\mathbf{n}(\mathbf{y}). \tag{10.55}$$

•

[10] Though the subset $\mathcal{V}_u$ of the set $\mathcal{V}$ from which the control $\mathbf{v}(\mathbf{u})$ is chosen in the case of a singular arc is *state dependent*, we have as a result of Lemma 10.5.4 that the optimal controls for the singular arc are on the boundary of $\mathcal{V}$ so that the argumentation of Theorem 5.6.1 is still valid.

Lemma 10.6.3

Let $D(\mathbf{x}) = 0$ represent a singular surface. If we define:

$$\overset{*}{H}^{(k)}(\mathbf{x}, J_x) = H^{(k)}[\mathbf{x}, J_x^{(k)}, \overset{*}{\mathbf{u}}(\mathbf{x}, J_x^{(k)}), \overset{*}{\mathbf{v}}(\mathbf{x}, J_x^{(k)})], \tag{10.56}$$

$$\overset{*}{\mathbf{f}}^{(k)}(\mathbf{x}, J_x) = \dot{\mathbf{x}}[\mathbf{x}, \overset{*}{\mathbf{u}}(\mathbf{x}, J_x^{(k)}), \overset{*}{\mathbf{v}}(\mathbf{x}, J_x^{(k)})], \tag{10.57}$$

we have:

$$\frac{\partial \overset{*}{H}^{(k)}}{\partial x} = \frac{\partial H^{(k)}}{\partial x}, \tag{10.58}$$

$$\frac{\partial \overset{*}{H}^{(k)}}{\partial J_x} = \frac{\partial H^{(k)}}{\partial J_x} = \overset{*}{\mathbf{f}}^{(k)}, \tag{10.59}$$

$$\frac{\partial^2 \overset{*}{H}^{(k)}}{\partial J_x^2} = \frac{\partial \overset{*}{\mathbf{f}}^{(k)}}{\partial J_x}. \tag{10.60}$$

Proof
The proof is left to the reader.

•

10.7 Properties of Regular Switch Surfaces

Switch surfaces are associated with cases where the velocity vector of the system is linear in component(s) of the controls of one or both players.
Let us assume, with no loss in generality, that φ is a component of the control $\mathbf{u}$ of player P and that $\mathbf{f}(\mathbf{x}, \mathbf{u}, \mathbf{v})$ is linear in φ.
A *switch surface*[11] for the control $\mathbf{u}$ is a locus of points of regular optimal trajectories where A_φ, the *switch function* for φ, vanishes. *Switch surfaces* separates two fields of regular optimal trajectories.
Optimal trajectories arrive from one side (denoted here as side-2) and leave to the other side (side-1).
We define a *regular switch surface* by the requirements:

1.
$$\begin{aligned} \overset{*}{\mathbf{u}}^{(1)} &\neq \overset{*}{\mathbf{u}}^{(2)}, \\ \overset{*}{\mathbf{v}}^{(1)} &= \overset{*}{\mathbf{v}}^{(2)} = \overset{*}{\mathbf{v}}, \end{aligned} \tag{10.61}$$

(we assumed, with no loss in generality that $\mathbf{u}$ switches).

[11] *R. Isaacs* named switch surfaces as *transition surface* and denoted them by *TS*.

2.
$$\begin{aligned}\mathbf{f}(\mathbf{x}, \overset{*}{\mathbf{u}}^{(1)}, \overset{*}{\mathbf{v}}) \cdot \mathbf{n}(\mathbf{x}) &< 0, \\ \mathbf{f}(\mathbf{x}, \overset{*}{\mathbf{u}}^{(2)}, \overset{*}{\mathbf{v}}) \cdot \mathbf{n}(\mathbf{x}) &< 0.\end{aligned} \tag{10.62}$$

Lemma 10.7.1

The function $\nabla J(\mathbf{x})$ is continuous across a regular *switch surface*.

Proof
Denote the *switch surface* as D. From Lemma 10.4.5 we know that if $\beta(\mathbf{x}) \neq 0$ the velocity vectors $\mathbf{f}(\mathbf{x}, \overset{*}{\mathbf{u}}^{(1)}, \overset{*}{\mathbf{v}}^{(2)})$ and $\mathbf{f}(\mathbf{x}, \overset{*}{\mathbf{u}}^{(2)}, \overset{*}{\mathbf{v}}^{(1)})$ do not lead to the same side of the surface D but this contradicts relation (10.62).

•

10.8 The Chatter Equivalent of Singular Arcs

10.8.1 Introduction

In section 10.5 we justified the use of a control law based on *nonadmissible information* of the type $\mathbf{u} = \mathbf{u}(\mathbf{v})$ along a *singular arc* for finding $J(\mathbf{x})$ by arguing that the singular arc is a limit of ϵ -optimal trajectories in the sense of the *extended solution concept* (see section 3.3).
We shall show[12] that the player, whom we assume apply a *nonadmissible control law* of the type $\mathbf{u} = \mathbf{u}(\mathbf{v})$ along a *singular arc*, has at his disposal *admissible* ϵ -optimal strategies that can guarantee him that the outcome of a play that starts at $\mathbf{x}$ will differ from the value $J(\mathbf{x})$ by no more than an arbitrarily chosen nonzero quantity.
Let us distinguish between two cases, the case where the *tributaries* join the singular arc *transversely* and the case where the junction is *tangential*.

10.8.2 Singular Arcs with Tributaries Joining Transversely

With no loss of generality we assume:

1. It is player P who is forced to use a *nonadmissible control law* of the type $\mathbf{u} = \mathbf{u}(\mathbf{v})$.

[12] We shall give the ideas of the proofs and skip some of the technical details.

2. The velocity vector $\mathbf{f}$ is linear in a component φ of the control vector $\mathbf{u}$ and $|\varphi| \leq \Phi$.

$$\mathbf{u} = \begin{pmatrix} u_p \\ \varphi \end{pmatrix},$$

$$\mathbf{f}(\mathbf{x}, \mathbf{u}, \mathbf{v}) = \mathbf{f_p}(\mathbf{x}, \mathbf{u_p}, \mathbf{v}) + \mathbf{b}(\mathbf{x})\varphi.$$

3. $A_\varphi(\mathbf{x}, J_x(\mathbf{x}))$ is the *switch function* for φ and

$$\overset{*}{\varphi} = -\text{sign} A_\varphi(\mathbf{x}, J_x(\mathbf{x})) \qquad \text{for} \quad A_\varphi(\mathbf{x}, J_x) \neq 0.$$

4. The tributaries to the singular arc arrive transversely at the singular surface D from side 1.

5. We assume that in region-1:

$$A_\varphi > 0 \ , \ \overset{*}{\varphi}^{(1)} = -\Phi.$$

Let us consider the following strategy of player P as a substitute for the nonadmissible control law $\overset{*}{\mathbf{u}}(\mathbf{v})$:

1. As the state arrives at $\mathbf{x} \in D$ player P plays

$$\tilde{\mathbf{u}}(\mathbf{x}) = \begin{pmatrix} \overset{*}{\mathbf{u}}_p(\mathbf{x}) \\ \Phi \end{pmatrix},$$

for a fixed duration δ[13].

2. Otherwise player P plays $\overset{*}{\mathbf{u}}^{(1)}(\mathbf{x})$ for states in side-1 of D and $\overset{*}{\mathbf{u}}^{(2)}(\mathbf{x})$ for states in side-2 of D[14].

Let us investigate the trajectory that corresponds to this strategy and the worse penalty in the outcome suffered by player P along it compared to the value that corresponds to the *singular arc*.
Suppose that the trajectory arrives (transversely) at the singular surface D from side-1 at the point o_1. Let us designate the time of arrival as $t = 0$. Player P now plays $\tilde{\mathbf{u}}(\mathbf{x})$ for a duration δ and the trajectory may lead back into side-1 and reach the point n_1 (see Figure 10.1).
Along the segment $o_1 n_1$ a loss is incurred upon player P because he plays nonoptimally.
From this instance player P resumes playing the control $\overset{*}{\mathbf{u}}^{(1)}(\mathbf{x})$ which is

[13] If the trajectory leads back to side-1 we have:

$$\mathbf{f}(\mathbf{x}, \tilde{\mathbf{u}}, \mathbf{v}) - \mathbf{f}(\mathbf{x}, \overset{*}{\mathbf{u}}^{(1)}, \mathbf{v}) = 2\mathbf{b}(\mathbf{x})\Phi.$$

[14] For the cases of *safe contact* a slight modification of the arguments is needed.

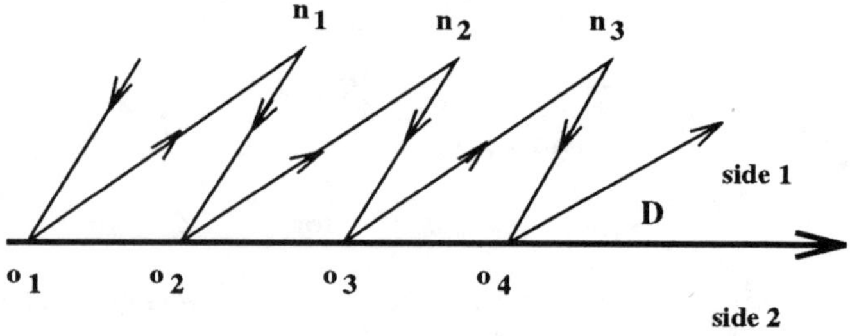

FIGURE 10.1. The chatter equivalent of a singular arc with transversal tributaries

optimal at side-1. The trajectory now leads *back* to the singular surface D and meets it transversely at the point o_2 and so forth.
We see that the trajectory *chatters* along the singular surface.
Let us estimate the increase $\Delta\pi_{o_1 n_1}$ in the outcome of a game from o_1 due to the possibly nonoptimal play by player P along the typical segment $o_1 n_1$.

$$\Delta\pi_{o_1 n_1} = J(\mathbf{x}_{n_1}) + \int_0^\delta L[\mathbf{x}(t), \tilde{\mathbf{u}}(\mathbf{x}(t)), \mathbf{v}(\mathbf{x}(t))]dt - J(\mathbf{x}_{o_1}). \tag{10.63}$$

Assuming a separable Hamiltonian and recalling the fact that region-1 is regular,

$$\Delta\pi_{o_1 n_1} = \int_0^\delta H[\mathbf{x}(t), J_x(\mathbf{x}(t)), \tilde{\mathbf{u}}(\mathbf{x}(t)), \mathbf{v}(\mathbf{x}(t))]dt. \tag{10.64}$$

in side-1:

$$\begin{aligned} H(\mathbf{x}, J_x(\mathbf{x}), \tilde{\mathbf{u}}(\mathbf{x}), \mathbf{v}) = \quad & H(\mathbf{x}, J_x(\mathbf{x}), \overset{*}{\mathbf{u}}(\mathbf{x}), \mathbf{v}) + 2\Phi A_\varphi(\mathbf{x}, J_x(\mathbf{x})) \leq \\ & 2\Phi A_\varphi(\mathbf{x}, J_x(\mathbf{x})). \end{aligned} \tag{10.65}$$

We assume that $\frac{dA_\varphi}{dt}$ is bounded in the vicinity of D by $\bar{A}$ and that player P can guarantee that the game from o_1 terminates in time less than T.
We have from Lemma 10.5.10 that A_φ vanishes on the regular optimal trajectories as they arrive from side 1 at the singular arc D.
This means that along the segment $o_1 n_1$

$$A_\varphi[\mathbf{x}(t), J_x(\mathbf{x}(t))] \leq \bar{A}t, \tag{10.66}$$

and

$$\Delta\pi_{o_1 n_1} \leq 2\Phi\bar{A}\delta^2, \tag{10.67}$$

and that the number $n(\delta)$ of periods of nonoptimal play by player P is bounded by the number $N(\delta)$

$$n(\delta) \leq N(\delta),$$

where

$$N(\delta)\delta = T. \tag{10.68}$$

Relation (10.68) implies that

$$\lim_{\delta \to 0} n(\delta)\delta^2 = \lim_{\delta \to 0} N(\delta)\delta^2 = 0,$$

so that

$$\lim_{\delta \to 0} \sum_{1}^{n(\delta)} \Delta\pi_{o,n_i} \leq \lim_{\delta \to 0} N(\delta)\Phi\bar{A}\delta^2 = 0,$$

which indicates that as $\delta \to 0$ the penalty that player P suffers due to his *chatter* strategy converges to 0. The described *chatter* strategy is hence an ϵ-optimal strategy.

10.8.3 Singular Arcs with Tributaries Joining Tangentially

With no loss of generality we assume:

1. It is player P who is forced to use a *nonadmissible control law* of the type $\mathbf{u} = \mathbf{u}(\mathbf{v})$.
2. The tributaries to the singular arc arrive at the singular surface D tangentially from side-1.
3. Let the point $\mathbf{x}$ be in the vicinity of the singular surface D in side-1 and let $\eta(\mathbf{x})$ be the distance of the point $\mathbf{x}$ from D. Player P has a control law $\tilde{\mathbf{u}}(\mathbf{x})$ that guarantees that $\frac{d\eta(\mathbf{x})}{dt} > b > 0$.
4. For any admissible $\mathbf{v}$ and time interval Δ the pair $(\overset{*}{\mathbf{u}}^{(1)}, \mathbf{v})$ provides
$$\left|\frac{\Delta\dot{\eta}(\mathbf{x})}{\Delta t}\right| < m.$$
5. Player P can guarantee that the game from points on the singular surface D terminates in time less than T.

Let us consider the following strategy of player P as a substitute for the nonadmissible control law $\overset{*}{\mathbf{u}}(\mathbf{v})$:

1. As the state arrives at $\mathbf{x} \in D$ player P plays $\tilde{\mathbf{u}}(\mathbf{x})$ for a fixed duration δ.

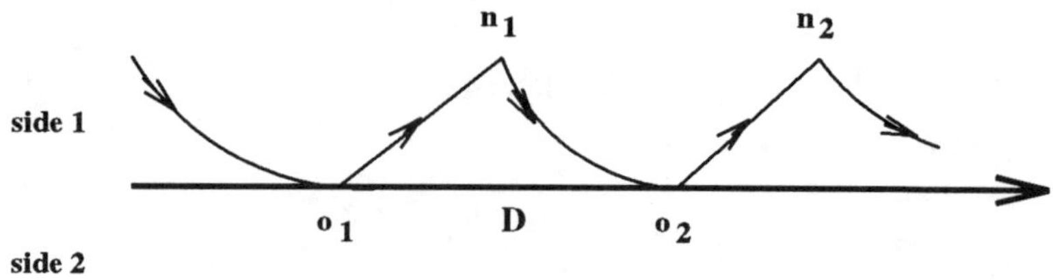

FIGURE 10.2. The chatter equivalent of a singular Arc with tangential Tributaries

2. Otherwise player P plays $\overset{*}{\mathbf{u}}^{(1)}(\mathbf{x})$ for states in side-1 of D and $\overset{*}{\mathbf{u}}^{(2)}(\mathbf{x})$ for states in side-2 of D[15].

Let us investigate the trajectory that corresponds to this strategy and the worse penalty in the outcome suffered by player P along it compared to the value that corresponds to the *singular arc*.
Suppose that the trajectory arrives (tangentially) at the singular surface D from side-1 at the point o_1. Let us designate the time of arrival as $t = 0$. Player P now plays $\tilde{\mathbf{u}}(\mathbf{x})$ for a duration δ and the trajectory may lead back into side-1 and reach the point n_1 (see Figure 10.2).
Along the segment $o_1 n_1$ a loss is incurred upon player P because he plays nonoptimally.

From this instance player P resumes playing the control $\overset{*}{\mathbf{u}}^{(1)}(\mathbf{x})$ which is optimal at side-1. The trajectory now leads *back* to the singular surface D and meets it *tangentially* at the point o_2. The duration of the segment $n_1 o_2$ is $\tilde{\delta}_1$.
Player P now plays $\tilde{\mathbf{u}}(\mathbf{x})$ for another duration δ and the trajectory may lead back into side-1 and reach the point n_2 and so forth.
We see that the trajectory *chatters* along the singular surface.
Let us estimate the total duration of the nonoptimal segments.
We have that:

$$\eta(\delta) \geq b\delta,$$

so the duration of a typical returning segment satisfies:

$$\tilde{\delta}_i \geq \sqrt{\frac{2\delta b}{m}}.$$

[15] For the cases of *safe contact* a slight modification of the arguments is needed.

The number $n(\delta)$ of periods of nonoptimal play by player P is bounded by the number $N(\delta)$

$$n(\delta) \leq N(\delta),$$

where

$$N(\delta)\sqrt{\frac{2\delta b}{m}} = T, \tag{10.69}$$

Relation (10.69) implies that

$$\lim_{\delta \to 0} n(\delta)\delta = \lim_{\delta \to 0} N(\delta)\delta = 0.$$

which means that as $\delta \to 0$ the total time spent along nonoptimal segments and the corresponding penalty that is incurred upon player P converge to 0. The *chatter* strategy is hence ϵ-optimal.

10.9 Sufficient conditions

In section 6.2 we described a proof of the *sufficiency theorem*, (Theorem 6.2.1), for the simple case when all the candidate optimal trajectories that cover the *capture-set* are regular. It is not difficult, in the context of the *extended value* concept, to extend the proof to the more general case when the capture set includes singular surfaces and the optimal trajectories include singular arcs segments. We do that by arguing that we can split the outcome-integral of the comparison plays $\pi[\mathbf{x}, \overset{*}{\mathbf{u}}(\cdot), v(\cdot)]$ and $\pi[\mathbf{x}, u(\cdot), \overset{*}{\mathbf{v}}(\cdot)]$ to finite number of parts corresponding to the regular parts and the chatter equivalent parts of the trajectories.

10.10 Problems

Problem 10.10.1

Study the case of a seam surface Q where:

$$\overset{*}{\mathbf{u}}^{(1)} = \overset{*}{\mathbf{u}}^{(2)} = \overset{*}{\mathbf{u}},$$

$$\mathbf{f}(\mathbf{x}, \overset{*}{\mathbf{u}}, \overset{*}{\mathbf{v}}^{(1)}) \cdot \mathbf{n}(\mathbf{x}) < 0, \tag{10.70}$$

$$\mathbf{f}(\mathbf{x}, \overset{*}{\mathbf{u}}, \overset{*}{\mathbf{v}}^{(2)}) \cdot \mathbf{n}(\mathbf{x}) > 0,$$

1. Show that for this case it is necessary that $\beta = 0$.
   Interpret the case.

•

Problem 10.10.2

Study the case where the regular optimal trajectories arrive at a seam surface Q from both sides and both controls $\mathbf{u}$ and $\mathbf{v}$ switch at the junction.

•

Problem 10.10.3

Study the case of a seam surface Q where:

$$\begin{aligned} &\overset{*}{\mathbf{u}}^{(1)} \neq \overset{*}{\mathbf{u}}^{(2)}, \\ &\mathbf{f}(\mathbf{x}, \overset{*}{\mathbf{u}}^{(1)}, \overset{*}{\mathbf{v}}^{(1)}) \cdot \mathbf{n}(\mathbf{x}) > 0, \\ &\mathbf{f}(\mathbf{x}, \overset{*}{\mathbf{u}}^{(2)}, \overset{*}{\mathbf{v}}^{(2)}) \cdot \mathbf{n}(\mathbf{x}) = 0. \end{aligned} \tag{10.71}$$

1. Show that this case is not feasible.

•

11

Dispersal Surfaces

11.1 introduction

In Example 7.3.3 (for the case when $V_m > U_m$) and in Example 7.3.4 we saw that the *regular* method for constructing regular optimal trajectories did not suffice for obtaining a *complete solution* and we had to conjecture the existence of a surface, (which we call, the way Isaacs did, a *dispersal surface*), separating two fields of trajectories that would otherwise intersect. We argued that when the game is at a point on such a *dispersal surface* one player can choose between two alternative controls yielding the same optimal outcome but leading to different sides.
We shall begin with discussion of the phenomenon of a region of multiple alternatives for the choices of controls (see Example 7.3.4).

11.2 Region of Multiple Choices

In the solutions of some game models we sometimes encounter regions where $\nabla J(\mathbf{x}) = 0$. This is analogous to a *plateau surface* in geographic topography.
It is easy to show (and it is left to the reader) that the boundary of such regions is necessarily semipermeable and that $\nabla J(\mathbf{x})$ may be discontinuous across it.
At the *plateau region* the choice of optimal controls is nonunique (as a matter of fact *all* choices are equally good!) so a multitude of optimal trajectories pass through each point.
Optimal trajectories in *plateau region* of $\nabla J(\mathbf{x}) = 0$ sometimes arrive at points or curves on the target-set (as in the solution to Problem 11.5.5) or at singular points or arcs inside the capture-set (as in Example 7.3.4).

11.3 Characterization of Dispersal Surfaces

A *dispersal surface* (to be denoted DS) is a *seam surface* that separates between two fields of regular optimal trajectories. Optimal trajectories *leave* the DS transversely to both sides.

$$
\begin{aligned}
&\mathbf{f}(\mathbf{x}, \overset{*}{\mathbf{u}}^{(1)}, \overset{*}{\mathbf{v}}^{(1)}) \cdot \mathbf{n}(\mathbf{x}) < 0, \\
&\mathbf{f}(\mathbf{x}, \overset{*}{\mathbf{u}}^{(2)}, \overset{*}{\mathbf{v}}^{(2)}) \cdot \mathbf{n}(\mathbf{x}) > 0.
\end{aligned}
\tag{11.1}
$$

Lemma 11.3.1

At a *dispersal surface*:

1.

$$\beta(\mathbf{x}) \neq 0,$$

2.

$$\overset{*}{\mathbf{u}}^{(1)}(\mathbf{x}) \neq \overset{*}{\mathbf{u}}^{(2)}(\mathbf{x}) , \qquad \overset{*}{\mathbf{v}}^{(1)}(\mathbf{x}) \neq \overset{*}{\mathbf{v}}^{(2)}(\mathbf{x}).$$

Proof
The proof is left to the reader.

•

Lemma 11.3.2

If $\beta(\mathbf{x}) > 0$ both $\overset{*}{\mathbf{v}}^{(1)}(\mathbf{x})$ (the optimal control of player E in field-1) and $\overset{*}{\mathbf{v}}^{(2)}(\mathbf{x})$ (the optimal control of player E in field-2) may fail to satisfy ME1 (equation 10.4.3) on the *dispersal surface*[1] .

Proof
As an immediate result of Lemma 10.4.5 we have:

$$
\begin{aligned}
&\mathbf{f}(\mathbf{x}, \overset{*}{\mathbf{u}}^{(1)}, \overset{*}{\mathbf{v}}^{(2)}) \cdot \mathbf{n}(\mathbf{x}) \leq 0, \\
&\mathbf{f}(\mathbf{x}, \overset{*}{\mathbf{u}}^{(2)}, \overset{*}{\mathbf{v}}^{(1)}) \cdot \mathbf{n}(\mathbf{x}) \geq 0.
\end{aligned}
\tag{11.2}
$$

At the seam between field-1 and field-2 (see Lemma 10.4.4) we have that:

$$H[\mathbf{x}, J_x^{(1)}, \overset{*}{\mathbf{u}}^{(1)}, \overset{*}{\mathbf{v}}^{(2)}] \leq 0, \tag{11.3}$$

$$H[\mathbf{x}, J_x^{(2)}, \overset{*}{\mathbf{u}}^{(1)}, \overset{*}{\mathbf{v}}^{(2)}] \geq 0, \tag{11.4}$$

$$H[\mathbf{x}, J_x^{(1)}, \overset{*}{\mathbf{u}}^{(2)}, \overset{*}{\mathbf{v}}^{(1)}] \geq 0, \tag{11.5}$$

---

[1]If $\beta(\mathbf{x}) < 0$ a similar assertion holds for the controls $\overset{*}{\mathbf{u}}^{(1)}(\mathbf{x})$ and $\overset{*}{\mathbf{u}}^{(2)}(\mathbf{x})$ of player P.

$$H[\mathbf{x}, J_x^{(2)}, \overset{*}{\mathbf{u}}^{(2)}, \overset{*}{\mathbf{v}}^{(1)}] \leq 0. \tag{11.6}$$

But Theorem 10.4.3 (ME1 on singular surfaces) requires that:

$$\begin{aligned} &H(\mathbf{x}, J_x^{(1)}, \mathbf{u}, \overset{*}{\mathbf{v}}) \geq 0\ , \quad \forall \mathbf{u} \in \mathcal{U} \text{ such that } \mathbf{f}(\mathbf{x}, \mathbf{u}, \overset{*}{\mathbf{v}}) \cdot \mathbf{n}(\mathbf{x}) \leq 0, \\ &H(\mathbf{x}, \mathbf{J}_x^{(2)}, \mathbf{u}, \overset{*}{\mathbf{v}}) \geq 0\ , \quad \forall \mathbf{u} \in \mathcal{U} \text{ such that } \mathbf{f}(\mathbf{x}, \mathbf{u}, \overset{*}{\mathbf{v}}) \cdot \mathbf{n}(\mathbf{x}) \geq 0. \end{aligned} \tag{11.7}$$

•

In the case of Lemma 11.3.2 we are led to the conclusion that at points $\mathbf{x}$ on a *dispersal surface* player E can guarantee an outcome equal to $J(\mathbf{x})$ only if he is allowed to use a *nonadmissible* control law of the type $\mathbf{v} = \mathbf{v}(\mathbf{u})$ such as:

$$\begin{aligned} &\overset{*}{\mathbf{v}}(\mathbf{u}) = \overset{*}{\mathbf{v}}^{(1)} \quad \text{if} \quad \mathbf{f}(\mathbf{x}, \mathbf{u}, \overset{*}{\mathbf{v}}^{(2)}) \cdot \mathbf{n}(\mathbf{x}) < 0, \\ &\overset{*}{\mathbf{v}}(\mathbf{u}) = \overset{*}{\mathbf{v}}^{(2)} \quad \text{if} \quad \mathbf{f}(\mathbf{x}, \mathbf{u}, \overset{*}{\mathbf{v}}^{(1)}) \cdot \mathbf{n}(\mathbf{x}) > 0. \end{aligned} \tag{11.8}$$

In this case player P dominates the choice into which side of the *DS* the game will follow.
We designate such a *DS* as a ***pursuer's dispersal surface*** (*PDS*).

R. Isaacs had already noticed that at a dispersal surface the strategy of one player may fail to satisfy the necessary conditions for optimality but he argued that this is only a *momentary dilemma* and that the loss (in outcome) incurred upon that player is infinitesimally small. He argued that as soon as the trajectory leaves the dispersal surface the regular optimal strategies for each side carry the optimal trajectories away from the dispersal surface. We leave it to the reader to show that we can replace the nonadmissible control law $\mathbf{v}(\mathbf{u})$ for states $\mathbf{x}$ that belong to the dispersal surface by ϵ-strategies that guarantee player E an outcome that differs from the value that corresponds to the use of the nonadmissible strategies by arbitrarily small $\epsilon > 0$.
It is also not difficult to modify the arguments of the *sufficiency theorem* (Theorem 6.2.1) and show that if the candidate composite solution (composed of regular fields of optimal trajectories seamed by dispersal surfaces) covers the capture-set it also qualifies as a *complete solution* in the sense of the extended solution concept.

11.4 Examples

Example 11.4.1 (The Surveillance Evasion game)

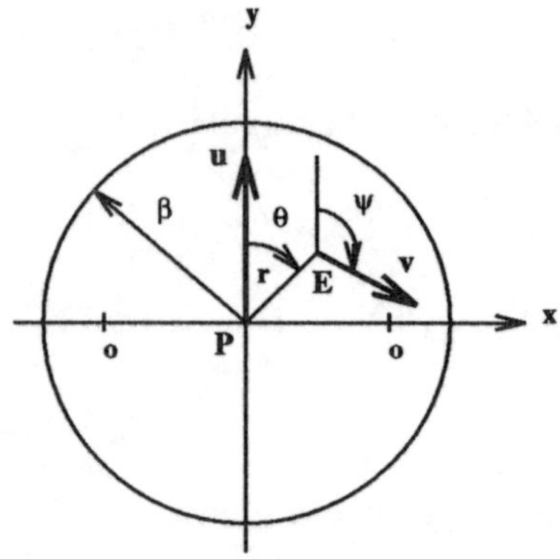

FIGURE 11.1. The Surveillance Evasion game

Let us consider the *Surveillance Evasion game*[2] .
A relatively fast ship P, with maximal speed $\sigma > 1$ and a minimal turn-radius of magnitude 1, tries to maintain a slower but more maneuverable vessel E, with maximal speed of magnitude 1 and with a capability of instantaneous turns, within a distance of magnitude β. (see Figure 11.1)
Let us formulate the game in a *relative frame* where the origin is attached to player P and the y-axis points in the direction of the velocity vector of player P.

The game-set:

$$\mathcal{S} \equiv \{(x,y) \mid x^2 + y^2 \leq \beta^2\} , \qquad \text{in a Cartesian frame.}$$

$$\mathcal{S} \equiv \{(r,\theta) \mid r \leq \beta\} , \qquad \text{in a polar frame.}$$

The target-set:

$$\mathcal{S} \equiv \{(x,y) \mid x^2 + y^2 = \beta^2\} , \qquad \text{in a Cartesian frame.}$$

$$\mathcal{S} \equiv \{(r,\theta) \mid r = \beta\} , \qquad \text{in a polar frame.}$$

The equations of motion:
In a Cartesian frame:

$$\begin{aligned} \dot{x} &= v \sin\psi - yu\varphi , \qquad |\varphi| \leq 1, \quad 0 \leq v \leq 1, \quad 0 \leq u \leq \sigma, \\ \dot{y} &= v \cos\psi - u + xu\varphi. \end{aligned} \tag{11.9}$$

[2]The problem was first discussed by *J. G. Taylor* (see [Tay70]) and the complete solution was given by *J. Lewin* and *J. V. Breakwell* in [LB75].

and in a polar frame:

$$\dot{r} = v\cos(\psi - \theta) - u\cos\theta\ , \qquad 0 \le v \le 1,\ \ 0 \le u \le \sigma,$$
$$r\dot{\theta} = v\sin(\psi - \theta) + u\sin\theta - ru\varphi\ , \quad |\varphi| \le 1. \tag{11.10}$$

Player P chooses u and φ and player E chooses v and ψ. $\sigma > 1$ and β are the parameters of the problem[3] .

The outcome functional:

$$G(x, y) = 0\ , \qquad L(x, y, u, \varphi, v, \psi) = 1.$$

In surveillance-evasion games it is the evader, player E, who prefers to terminate and to minimize the time-to-escape. Player P prefers nontermination over any numerical outcome.

Solution

Partitioning of the target-set
Let $\hat{\mathbf{e}}_r$ be a unit vector in the r-direction. The *normal* to the target-set at (β, θ) is

$$\mathbf{n}(\beta, \theta) = -\hat{\mathbf{e}}_r.$$

We have:

$$\max_{u} \min_{v,\sigma}[u\cos\theta - v\cos(\psi - \theta)] = \sigma\cos\theta - 1.$$

so:

$$\mathrm{UP} \equiv \{(\beta, \theta) \mid (1 - \sigma\cos\theta) > 0\},$$

$$\mathrm{NUP} \equiv \{(\beta, \theta) \mid (1 - \sigma\cos\theta) < 0\},$$

$$\mathrm{BUP} \equiv \{(\beta, \theta) \mid (1 - \sigma\cos\theta) = 0\}.$$

Candidate optimal control laws
The equations of motion are *separable* so we can express the Isaacs equation ME1 in the *Cartesian frame* as:

$$\max_{u,\varphi} \min_{v,\psi}[(v\sin\psi - yu\varphi)J_x + (v\cos\psi - u + xu\varphi)J_y + 1],$$

[3]This is a normalized version and the two parameters of the game σ and β can be regarded as the ratios between the speeds of the pursuer and the evader and the ratio between the radius of the surveillance circle and the minimal radius of turn respectively.

and in the *polar* frame:

$$\max_{u,\varphi}\min_{v,\psi}\{[v\cos(\psi-\theta)-u\cos\theta]J_r+[v\sin(\psi-\theta)+u\sin\theta-u\varphi]\tfrac{J_\theta}{r}+1\}.$$

Let $A_c(x,y)$ be the *switch function* for $\overset{*}{\varphi}$ in the Cartesian representation.

$$A_c(x,y)=xJ_y(x,y)-yJ_x(x,y).$$

The candidate optimal control laws in the Cartesian frame for player E are:

$$\begin{gathered}\overset{*}{v}=1\\(\sin\overset{*}{\psi},\cos\overset{*}{\psi})\parallel(-J_x,-J_y).\end{gathered}\tag{11.11}$$

and for player P:

$$\begin{aligned}\overset{*}{\varphi}&=\operatorname{sign}A_c(x,y)\ , && \text{for } A_c(x,y)\neq 0,\\ \overset{*}{\varphi}&=\text{any admissible }\varphi\ , && \text{for } A_c(x,y)=0,\\ \\ \overset{*}{u}&=0\ , && \text{for } |A_c(x,y)|-J_y(x,y)\leq 0,\\ \overset{*}{u}&=\sigma\ , && \text{for } |A_c(x,y)|-J_y(x,y)>0.\end{aligned}\tag{11.12}$$

Similarly for the representation in polar coordinates we have that $A_p(x,y)$, the *switch function* for $\overset{*}{\varphi}$, is

$$A_p(r,\theta)=J_\theta(r,\theta).$$

the candidate optimal control laws for player E are:

$$\begin{gathered}\overset{*}{v}=1,\\(\cos(\overset{*}{\psi}-\theta),(\sin(\overset{*}{\psi}-\theta))\parallel(-J_r,-J_\theta),\end{gathered}\tag{11.13}$$

and for player P:

$$\begin{aligned}\overset{*}{\varphi}&=\operatorname{sign}A_p(r,\theta)\ , && \text{for } A_p(r,\theta)\neq 0,\\ \overset{*}{\varphi}&=\text{any admissible }\varphi\ , && \text{for } A_p(r,\theta)=0,\\ \\ \overset{*}{u}&=0\ , && \text{for } \tfrac{1}{r}J_\theta\sin\theta-J_r\cos\theta+J_\theta\leq 0,\\ \overset{*}{u}&=\sigma\ , && \text{for } \tfrac{1}{r}J_\theta\sin\theta-J_r\cos\theta+J_\theta>0.\end{aligned}\tag{11.14}$$

Notice that the switch in $\overset{*}{\varphi}$ takes place when $\nabla J(x,y)$ passes through P and that the switch in $\overset{*}{u}$ occurs when $\nabla J(x,y)$ passes through Q or Q' (the centers for right turn or left turn respectively).

The adjoint equations

$$
\begin{aligned}
\overset{\circ}{J}_x\,[x(\tau), y(\tau)] &= J_y[x(\tau), y(\tau)]\,\overset{*}{u}\overset{*}{\varphi},\\
\overset{\circ}{J}_y\,[x(\tau), y(\tau)] &= -J_x[x(\tau), y(\tau)]\,\overset{*}{u}\overset{*}{\varphi},
\end{aligned}
\tag{11.15}
$$

and this provides:

$$
J_x[x(\tau), y(\tau)]\,\overset{\circ}{J}_x\,[x(\tau), y(\tau)] + J_y[x(\tau), y(\tau)]\,\overset{\circ}{J}_y\,[x(\tau), y(\tau)] = 0, \tag{11.16}
$$

Which means that $\nabla J(x, y)$ is constant along regular segments of optimal trajectories. We also have:

$$
\overset{\circ}{A}\,(x, y) = -\,\overset{*}{u}\,J_x(x, y). \tag{11.17}
$$

Terminal values of $\nabla J(x, y)$
The tangent $\mathbf{t}(\beta, \theta)$ to the target-set at (β, θ) is $\mathbf{t} = \hat{\mathbf{e}}_\theta$ in the polar representation and $\mathbf{t} = y\hat{\mathbf{e}}_x - x\hat{\mathbf{e}}_y$ in the Cartesian frame.
We have that $\nabla G(r, \theta) = 0$ so:

$$
J_\theta(\beta, \theta) = 0.
$$

From ME-2 we obtain in the polar case:

$$
-|J_r(\beta, \theta)| - J_r(\beta, \theta)\,\overset{*}{u}\,\cos\theta + 1 = 0.
$$

The requirement that player E guarantees termination at the points of the UP requires that $\cos(\overset{*}{\psi} - \theta) \geq 0$ so that $J_r(\beta, \theta) \leq 0$. We obtain:

$$
\begin{aligned}
J_r(\beta, \theta) &= -1/(1 - \sigma\cos\theta)\,, && \text{for} \quad \cos^{-1}\tfrac{1}{\sigma} < |\theta| < \tfrac{\pi}{2},\\
J_r(\beta, \theta) &= -1\,, && \text{for} \quad \tfrac{\pi}{2} < |\theta| \leq \pi,\\
|\nabla J(\beta, \theta)| &= 1/(1 - \sigma\cos\theta)\,, && \text{for} \quad \cos^{-1}\tfrac{1}{\sigma} < |\theta| < \tfrac{\pi}{2},\\
|\nabla J(\beta, \theta)| &= 1\,, && \text{for} \quad \tfrac{\pi}{2} < |\theta| \leq \pi,
\end{aligned}
\tag{11.18}
$$

and in the Cartesian reference frame:

$$
\begin{aligned}
J_x(\beta\sin\theta, \beta\cos\theta) &= -\sin\theta/(1 - \sigma\cos\theta)\,, && \text{for} \quad \cos^{-1}\tfrac{1}{\sigma} < |\theta| < \tfrac{\pi}{2},\\
J_x(\beta\sin\theta, \beta\cos\theta) &= -\sin\theta\,, && \text{for} \quad \tfrac{\pi}{2} < |\theta| \leq \pi,\\
J_y(\beta\sin\theta, \beta\cos\theta) &= -\cos\theta/(1 - \sigma\cos\theta)\,, && \text{for} \quad \cos^{-1}\tfrac{1}{\sigma} < |\theta| < \tfrac{\pi}{2},\\
J_y(\beta\sin\theta, \beta\cos\theta) &= -\cos\theta\,, && \text{for} \quad \tfrac{\pi}{2} < |\theta| \leq \pi.
\end{aligned}
\tag{11.19}
$$

The candidate optimal trajectories

Let us parametrize the domain $\cos^{-1}\frac{1}{\sigma} \leq \theta \leq \pi/2$ of the UP by the parameter θ:

$$x_0 = \beta \sin\theta \ , \quad y_0 = \beta\cos\theta.$$

The first field (field-1) of optimal trajectories is obtained by solving the following system of equations from points on the UP.

$$\begin{aligned}
\overset{\circ}{J_x}[x,y] &= \sigma \overset{*}{\varphi} J_y[x,y] \ , & J_x[x(0),y(0)] &= -\sin\theta/(1-\sigma\cos\theta),\\
\overset{\circ}{J_y}[x,y] &= -\sigma \overset{*}{\varphi} J_x[x,y] \ , & J_y[x(0),y(0)] &= -\cos\theta/(1-\sigma\cos\theta),\\
\overset{\circ}{A}(x,y) &= -\sigma J_x(x,y) \ , & A[x(0),y(0)] &= 0.
\end{aligned} \tag{11.20}$$

We obtain:

$$\begin{aligned}
J_x[x(\tau),y(\tau)] &= -\sin(\theta+\sigma\overset{*}{\varphi}\tau)/(1-\sigma\cos\theta),\\
J_y[x(\tau),y(\tau)] &= -\cos(\theta+\sigma\overset{*}{\varphi}\tau)/(1-\sigma\cos\theta),\\
A[x(\tau),y(\tau)] &= [\cos(\theta+\sigma\overset{*}{\varphi}\tau)-\cos\tau]/[(1-\sigma\cos\theta)\overset{*}{\varphi}].
\end{aligned} \tag{11.21}$$

A switch may occur at $\tau = 2(\pi-\tau)/\sigma$.
We get the candidate optimal trajectories by solving

$$\begin{aligned}
\overset{\circ}{x}(\tau) &= -\sin(\theta+\sigma\overset{*}{\varphi}\tau)+y\sigma\overset{*}{\varphi} \ , & x(0) &= \beta\sin\theta,\\
\overset{\circ}{y}(\tau) &= -\cos(\theta+\sigma\overset{*}{\varphi}\tau)+\sigma(1-x\overset{*}{\varphi}) \ , & y(0) &= \beta\cos\theta.
\end{aligned} \tag{11.22}$$

We obtain

$$\begin{aligned}
x(\tau) &= (\beta-\theta)\sin(\theta+\sigma\overset{*}{\varphi}\tau)+\frac{1}{\overset{*}{\varphi}}[1-\cos(\theta+\sigma\overset{*}{\varphi}\tau)],\\
y(\tau) &= (\beta-\theta)\cos(\theta+\sigma\overset{*}{\varphi}\tau)+\frac{1}{\overset{*}{varphi}}[\sin(\theta+\sigma\overset{*}{\varphi}\tau)].
\end{aligned} \tag{11.23}$$

Another field of optimal trajectories (field-2) is obtained for $\frac{\pi}{2} < |\theta| \leq \pi$. We have $\overset{*}{u} = 0$ and we have to solve:

$$\begin{aligned}
\overset{\circ}{J_x}(x,y) &= 0 \ , & J_x[x(0),y(0)] &= -\sin\theta/(1-\sigma\cos\theta),\\
\overset{\circ}{J_y}(x,y) &= 0 \ , & J_y[x(0),y(0)] &= -\cos\theta/(1-\sigma\cos\theta).\\
\overset{\circ}{A}(x,y) &= 0 \ , & A[x(0),y(0)] &= 0.
\end{aligned} \tag{11.24}$$

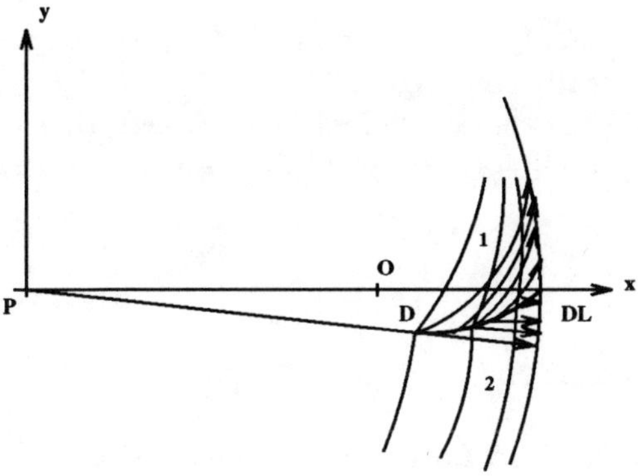

FIGURE 11.2. Isochrones for the Surveillance Evasion game

We get

$$J_x[x(\tau), y(\tau)] = -\sin\theta/(1 - \sigma\cos\theta),$$

$$J_y[x(\tau), y(\tau)] = -\cos\theta/(1 - \sigma\cos\theta), \tag{11.25}$$

$$A[x(\tau), y(\tau)] = 0.$$

And by solving:

$$\overset{\circ}{x}(\tau) = -\sin\theta\ , \quad x(0) = \beta\sin\theta,$$
$$\overset{\circ}{y}(\tau) = -\cos\theta\ , \quad x(0) = \beta\cos\theta, \tag{11.26}$$

we obtain the candidate optimal travjectories for field-2:

$$x(\tau) = (\beta - \tau)\sin\theta,$$
$$y(\tau) = (\beta - \tau)\cos\theta. \tag{11.27}$$

The candidate optimal trajectories in the vicinity of $\theta = \frac{1}{2}\pi$[4] for

$$1 < \beta < f_1(\sigma),$$

where

$$f_1(\sigma) = \sqrt{1 - \frac{1}{\sigma^2}} + \frac{1}{\sigma}(1 + \frac{1}{2}\pi - 2\sin^{-1}\frac{1}{\sigma}),$$

are shown in Figure 11.2.
Consider the second quadrant. The point where a τ-isochrone of field-1 intersects the τ-isochrone of field-2 is a candidate dispersal-point. We can

[4] An *open barrier* emanates from the *BUP*. For more details see [LB75]

construct the locus of those points, (the *dispersal surface*), either geometrically or analytically (see Figure 11.2).
Notice that the construction cannot be carried further back from the point D where the optimal trajectories of field-2 join the *dispersal surface* tangentially.

•

11.5 Problems

Problem 11.5.1 (The Lady in the Lake) *[Isa75]*

Consider Example 7.3.4.

1. Show that the pursuer dominates the *dispersal line*.
2. Find an expresion for $\tilde{\theta}$. (where the optimal trajectory that emanates from $(1, \tilde{\theta})$ reaches $(\beta, 0)$).
3. Examine the corner between the two semipermeable surfaces formed by the optimal trajectories that emanate from $(1, \tilde{\theta})$ and $(1, -\tilde{\theta})$

•

Problem 11.5.2 (The Surveillance Evasion game) *[LB75]*

Consider Example 11.4.1.

1. Find an expression for the slope of the *dispersal surface*.
2. Suggest a way to construct the *dispersal surface* geometrically. (Use the geometrical characterization of the iso-value surfaces).
3. Show that the *dispersal surface* is a *PDL*.

•

Problem 11.5.3

Consider Example 7.3.3.
Determine which player dominates the *dispersal surface*.

•

Problem 11.5.4

In a case of a *PDL* player P can use a *chatter* control law to stay adjacent to the *PDL*.

1. Write a necessary condition that states that player P cannot benefit from this option.

•

Problem 11.5.5

Consider Example 7.3.1 but with a different terminal cost:

$$G(x, y) = -x^2$$

1. Solve the game.

•

Problem 11.5.6

Consider *pure pursuit evasion games* $(G = 0, \quad L = 1)$.

1. Show that in *pure pursuit evasion games* it is not feasible to have a *plateau* in the value map.

•

Problem 11.5.7

Consider the following problem:

The game-set:

$$\mathcal{S} \equiv \{(x, y) \mid \; x^2 + y^2 \geq r^2\}.$$

The capture-set:

$$\mathcal{S} \equiv \{(x, y) \mid \; x^2 + y^2 = r^2\}.$$

The equations of motion:

$$\dot{x} = u_1 + v \quad , \; |v| \leq 1,$$

$$\dot{y} = u_2 \quad , \; u_1^2 + u_2^2 = w^2 \quad , \; 0 < w < 1.$$

The outcome functional:

$$G(x, y, u_1, u_2, v) = 0 \quad , \; L(x, y, u_1, u_2, v) = 1.$$

1. Solve for the primary field of optimal trajectories.
2. Show that all the trajectories pass through $(\frac{r}{w}, 0)$ or $(-\frac{r}{w}, 0)$.

3. Examine the candidate *dispersal surface* (on the y-axis) and determine which player dominates it.

4. Show that the paths that emanate from the BUP form a *closed barrier*.

5. Derive an expression for the value function. Draw the map of equal-value-surfaces for the case of $w = \frac{1}{2}$.

•

12

Safe Contact

12.1 Introduction

In some real pursuit-evasion situations the pursuer has a *speed* advantage while the evader is more *maneuverable*. The evader may sometimes be better off if for a while he uses a *brinkmanship* tactics that maintains the motion dangerously close to being captured by the pursuer in order to realize his superior maneuverability.

A similar phenomenon, which we call *safe contact*, may occur in differential games. An optimal trajectory may head to the *NUP* of the target-set or to a *barrier* and then adhere to them in the form of a *singular arc* for a while.

In this chapter we shall characterize this type of *singular arcs* and discuss the methods for their construction .
We recall that in order to realize the *safe contact* part of an optimal trajectory and the corresponding outcome $J(\mathbf{x})$ one player must base his choice of controls upon prior (nonadmissible!) knowledge of his opponent's choice of control. We also remember (see section 10.8) that this player has at his disposal *admissible* ϵ -optimal strategies that can guarantee him that the outcome of a play that starts at $\mathbf{x}$ on the *safe contact* surface will differ from the value $J(\mathbf{x})$ by no more than an arbitrarily chosen nonzero quantity.
We shall utilize the necessary conditions regarding the junction of regular optimal trajectories arriving at a singular arc (that where discussed in section 10.5) to provide us with information that is necessary for the backwards construction of regular optimal trajectories (tributaries) that join the *safe contact* arcs.

12.2 Characterization of Safe Contact

Safe Contact takes place adjacent to side-i of a *barrier* surface or adjacent to the *NUP*. Regular optimal trajectories of field-i *arrive* at side-i of the *barrier* or at the *NUP* and proceed adjacent to it as *singular arcs*.
Let $\mathbf{x}$ be a point on the *safe contact* surface Q and let $\mathbf{n}(\mathbf{x})$ be a vector normal to the *safe contact* surface Q at $\mathbf{x}$ pointing into side-i. At $\mathbf{x}$ we have:

1.
$$\mathbf{f}(\mathbf{x}, \overset{*}{\mathbf{u}}^{(i)}, \overset{*}{\mathbf{v}}^{(i)}) \cdot \mathbf{n}(\mathbf{x}) \leq 0. \tag{12.1}$$

2. If player E is allowed to realize a nonadmissible control law of the type $\mathbf{v}(\mathbf{u})$ then there exist control laws $\tilde{\mathbf{v}}(\mathbf{u})$ such that,
$$\mathbf{f}(\mathbf{x}, \mathbf{u}, \tilde{\mathbf{v}}(\mathbf{u})) \cdot \mathbf{n}(\mathbf{x}) \geq \mathbf{0}\ , \quad \text{for all } \ \mathbf{u} \in \mathcal{U}, \tag{12.2}$$
and specifically:
$$\mathbf{f}(\mathbf{x}, \overset{*}{\mathbf{u}}^{(i)}, \tilde{\mathbf{v}}(\overset{*}{\mathbf{u}}^{(i)})) \cdot \mathbf{n}(\mathbf{x}) = 0. \tag{12.3}$$

Relations (12.1) and (12.2) indicate that the *safe contact* falls into the category of *singular arcs* discussed previously (see section 10.5) and that there are two kinds of junctions of the incoming *tributaries* (the regular optimal trajectories of field-i) to the safe contact surface Q.

1. A *tangential* junction is associated with the case when (see *junction condition 2*, Theorem 10.5.8) the maximization problem:
$$\max_{\mathbf{v} \in \mathcal{V}} H[\mathbf{x}, \mathbf{J}_{\mathbf{x}}^{(\mathbf{i})}, \overset{*}{\mathbf{u}}^{(\mathbf{i})}, \mathbf{v}], \tag{12.4}$$
has a *unique solution*.

2. A *transversal* junction happens in the cases (see *junction condition 4*, Theorem 10.5.10) when the maximization problem:
$$\max_{\mathbf{v} \in \mathcal{V}} H[\mathbf{x}, \mathbf{J}_{\mathbf{x}}^{(\mathbf{i})}, \overset{*}{\mathbf{u}}^{(\mathbf{i})}, \mathbf{v}], \tag{12.5}$$
has a *nonunique* solution.

   (This happens when the Hamiltonian $H[\mathbf{x}, \mathbf{J}_{\mathbf{x}}^{(\mathbf{i})}, \overset{*}{\mathbf{u}}^{(\mathbf{i})}, \mathbf{v}]$ is linear in component(s) of $\mathbf{v}$).

12.3 Construction of Safe Contact Arcs

12.3.1 Introduction

When we solve a differential game we usually do not know, ahead of time, if the optimal trajectories have singular segments or not. Only when the regular backward construction of candidate optimal trajectories fails to cover the game-set [1] we have to assume the existence of a *singular surface*

[1] If an *escape-set*, bounded by *closed barriers* was identified then we need only to cover the *capture-set*.

(which is a manifold of *singular arcs*) and of manifolds of tributaries joining it, in order to proceed.
The specific singular phenomenon depends on the character of the the specific problem that we solve. Some singular phenomena are not easy to identify. Luckily, the clues for the possible existence of *safe contact* are simple:

1. Relation (12.2) has to be satisfied on a candidate *safe contact surface* Q (usually on part of a *barrier* or on part of the *NUP*).
2. A curve S that belong to a previously constructed field of candidate optimal trajectories also belongs to the surface Q.

We start from points that belong the the curve S and we construct *safe contact* arcs that lie on the *safe contact surface* by using *Isaacs equations* and by integrating the *adjoint equations for singular arcs*[2] (see relation (10.44)):

$$\frac{d}{dt}\nabla J^{(k)}(\mathbf{y}) = -\frac{\partial H(\mathbf{y}, \mathbf{J}_{\mathbf{y}}^{(\mathbf{k})}, \overset{*}{\mathbf{u}}, \overset{*}{\mathbf{v}})}{\partial \mathbf{y}} + \Gamma(t)\mathbf{n}(\mathbf{y}), \tag{12.6}$$

where $\mathbf{n}(\mathbf{y})$ is the normal to the candidate *safe contact surface* Q.
Let us follow *P. Bernhard* (see [Ber77]) and use the *junction conditions* to derive an equation for computing $\Gamma(t)$[3].

Let us denote the velocity vector and the value of the Hamiltonian on the incoming tributaries at the point $\mathbf{x}$ where they join the *safe contact* surface Q by $\overset{*}{\mathbf{f}}$ and $\overset{*}{H}$ respectively (see Lemma 10.6.3).

12.3.2 Safe Contact with Tangential Junctions

Let $Q(\mathbf{x}) = 0$ represent the *safe contact* surface. We may compute the field of singular arcs on the *safe contact* surface by integrating :

$$\dot{\mathbf{x}} = \overset{*}{\mathbf{f}}, \tag{12.7}$$

$$\dot{\mathbf{J}}_x = -\frac{\partial \overset{*}{H}}{\partial x} + \Gamma(t)\frac{\partial Q}{\partial x}, \tag{12.8}$$

---

[2] In 2-dimensional games it often suffices to use ME2 at the adjacent region in order to find the magnitude of the adjoint vector at the junction with the tributaries.

[3] In some cases the *safe contact* arcs can be considered as regular optimal trajectories of the solution of a reduced order game constrained to takes place on the *safe contact* surface Q. The *target-set* of this reduced-game is a (n-2)-dimensional curve S that borders the region where a candidate value function $J(\mathbf{x})$ is already available. (Those values will serve as the *terminal cost function* for the reduced game). We must remember that this is not always the case (see Problem 12.5.4).

subject to the requirement that:

$$Q_x \overset{*}{\mathbf{f}} = 0. \tag{12.9}$$

We may find $\Gamma(t)$ along the singular arcs by differentiating relation (12.9) with respect to time. It is not difficult to show that this results in:

$$\overset{*}{\mathbf{f}}^T Q_{xx} \overset{*}{\mathbf{f}} + Q_x \overset{*}{\mathbf{f}}_x \overset{*}{\mathbf{f}} + Q_x \overset{*}{\mathbf{f}}_{J_x} (\dot{\mathbf{J}}_x)^T = 0. \tag{12.10}$$

Substituting from relation (12.8) we obtain the desired equation for $\Gamma(t)$:

$$\overset{*}{\mathbf{f}}^T Q_{xx} \overset{*}{\mathbf{f}} + Q_x \overset{*}{\mathbf{f}}_x \overset{*}{\mathbf{f}} - Q_x \overset{*}{\mathbf{f}}_{J_x} \left(\frac{\partial \overset{*}{H}}{\partial x}\right)^T + \Gamma(t) Q_x \overset{*}{\mathbf{f}}_{J_x} (Q_x)^T = 0. \tag{12.11}$$

12.3.3 Safe Contact with Transversal Junctions

Let us define the switch function $\overset{*}{a}(\mathbf{x}, \mathbf{J}_x)$ as

$$\overset{*}{a}(\mathbf{x}, J_x) = a(\mathbf{x}, \mathbf{J}_x, \overset{*}{\mathbf{u}}(\mathbf{x}, _x), \overset{*}{\mathbf{v}}(\mathbf{x}, \mathbf{J}_x)), \tag{12.12}$$

where $\overset{*}{a}$ vanishes on the tributaries as they join the singular arcs.
We may construct the field of singular arcs on the *safe contact* surface by integrating the equations (12.7) and (12.8) subject to the requirement

$$\overset{*}{a}(\mathbf{x}, \mathbf{J}_x) = 0. \tag{12.13}$$

Differentiating relation (12.13) with respect to time enables us to find $\Gamma(t)$ in the course of the construction. It is not difficult to show that this gives:

$$\overset{*}{a}_x \overset{*}{\mathbf{f}} + \overset{*}{a}_{J_x} (\dot{\mathbf{J}}_x)^T = 0. \tag{12.14}$$

substituting from relation (12.8) we obtain:

$$a_x \overset{*}{\mathbf{f}} - a_{J_x} \left(\frac{\partial \overset{*}{H}}{\partial x}\right)^T + \Gamma(t) a_{J_x} (Q_x)^T = 0. \tag{12.15}$$

As we construct the singular arc we can also construct the tributary field emanating backwards from points on the singular arc into the game-set thus filling it with an additional field of regular optimal trajectories.

12.4 Examples

Example 12.4.1 (The Dolichobrachistochrone)

Let us recall Example 5.2.1 and study it further.

The game-set:
The game takes place in a quadrant of a plane.

$$\mathcal{S} \equiv \{(x,y) \mid x \geq 0,\ y \geq 0\}.$$

The target-set:

$$\mathcal{C} \equiv \{(x,y) \mid x = 0,\ y \geq 0\}.$$

The equations of motion:

$$\dot{x} = \sqrt{y}\cos\varphi + \tfrac{1}{2}w(\psi + 1)\ , \quad |\psi| \leq 1,$$

$$\dot{y} = \sqrt{y}\sin\varphi + \tfrac{1}{2}w(\psi - 1).$$

Player P chooses φ and player E chooses ψ.

The outcome functional:
The game is a pure pursuit evasion game:

$$L(x,y,\varphi,\psi) = 1\ , \quad G(x,y) = 0.$$

Solution

The partitioning of the target-set
It is easy to see that the choice of $\tilde{\psi} = 1$ guarantees player E that no play will terminate at points on the target-set where $y < w^2$. So

$$\text{NUP} \equiv \left\{(x,y) \mid x = 0,\ y < w^2\right\},$$

$$\text{UP} \equiv \left\{(x,y) \mid x = 0,\ y > w^2\right\},$$

$$\text{BUP} \equiv \left\{(x,y) \mid x = 0,\ y = w^2\right\}.$$

The candidate optimal control laws
The equations of motion are separable so we can state the Isaacs equations as:

$$\min_{\varphi}\max_{|\psi|\leq 1}\{J_x[\sqrt{y}\cos\varphi+\frac{1}{2}w(\psi+1)]+J_y[\sqrt{y}\sin\varphi+\frac{1}{2}w(\psi-1)]+1\}. \quad (12.16)$$

Let us define the switch function for the control $\overset{*}{\psi}$ and the function $\sigma(x,y)$ as:

$$A(x,y) = J_x(x,y) + J_x(x,y)\ , \qquad \sigma(x,y) = |\nabla J(x,y)|. \qquad (12.17)$$

The candidate laws for the choice of the optimal control $\overset{*}{\psi}$ by player E are:

$$\begin{aligned} \overset{*}{\psi} &= \text{sign } A(x,y)\ , && \text{for } A(x,y) \neq 0, \\ \overset{*}{\psi} &= \text{any admissible } \psi\ , && \text{for } A(x,y) = 0. \end{aligned} \tag{12.18}$$

and the candidate law for the choice of the optimal control $\overset{*}{\varphi}$ by player P:

$$(\cos \overset{*}{\varphi}(x,y), \sin \overset{*}{\varphi}(x,y)) \parallel (-J_x(x,y), -J_y(x,y)). \tag{12.19}$$

The adjoint equations

$$\begin{aligned} \overset{\circ}{J}_x\, [x(\tau), y(\tau)] &= 0, \\ \overset{\circ}{J}_y\, [x(\tau), y(\tau)] &= \tfrac{\sigma}{2\sqrt{y}}. \end{aligned} \tag{12.20}$$

Adjacent to the target-set, where $A(x,y) > 0$, we can obtain by ME2:

$$\overset{\circ}{J}_y\, [x(\tau), y(\tau)] = \overset{\circ}{A}\, [x(\tau), y(\tau)] = \frac{\sigma^2}{2(1 + wJ_x[x(\tau), y(\tau)]}. \tag{12.21}$$

The terminal values of $\nabla J(\mathbf{x})$

The tangent to the target-set at $(0,y)$ is spanned by $\hat{\mathbf{e}}_y$ (a unit vector in the y-direction). Also $\nabla G(x,y) = 0$.

At points on the *UP* where optimal trajectories terminate we have that in order to guarantee termination $\cos \overset{*}{\varphi} < 0$. By equations (12.18) $J_x(0,y) > 0$ and hence $A(0,y) > 0$.

Using ME-2 we get:

$$\begin{aligned} J_x(0,y) &= 1/(\sqrt{y} - w), \\ J_y(0,y) &= 0. \end{aligned} \tag{12.22}$$

The first field of candidate optimal trajectories

It is not difficult to obtain that:

$$J_x(x,y) = \frac{-1}{w - \sqrt{y_0}}, \tag{12.23}$$

$$J_y(x,y) = \frac{\sqrt{\frac{y_0}{y} - 1}}{w - \sqrt{y_0}}, \tag{12.24}$$

$$\sigma(x,y) = -\frac{\sqrt{\frac{y_0}{y}}}{w - \sqrt{y_0}}. \tag{12.25}$$

The equations for the optimal trajectories are:

$$\dot{x} = w - \frac{y}{\sqrt{y_0}}, \tag{12.26}$$

$$\dot{y} = \sqrt{y - \frac{y^2}{y_0}}. \tag{12.27}$$

The candidate optimal trajectory from a point $(0, y_0)$ on the *UP* is given by:

$$\begin{aligned} x(\tau) &= \tfrac{1}{2}\tau\sqrt{y_0} + \tfrac{1}{2}y_0 \sin\tfrac{\tau}{\sqrt{y_0}} - w\tau, \\ y(\tau) &= \tfrac{1}{2}y_0(1 + \cos\tfrac{\tau}{\sqrt{y_0}}). \end{aligned} \tag{12.28}$$

The switch function $A(x, y)$ for $\overset{*}{\psi}$ is given by:

$$A(x, y) = \frac{1 - \sqrt{\frac{y_0}{y} - 1}}{w - \sqrt{y_0}}. \tag{12.29}$$

The slope of the candidate optimal trajectory is:

$$\frac{dx}{dy} = \frac{w\sqrt{y_0} - y}{\sqrt{y}\sqrt{y_0 - y}}. \tag{12.30}$$

so the optimal trajectory is given by:

$$x = (\frac{1}{2}y_0 - w\sqrt{y_0}) \cos^{-1}\frac{2y - y_0}{y_0} + \sqrt{y}\sqrt{y_0 - y}. \tag{12.31}$$

A switch from $\overset{*}{\psi}= 1$ to $\overset{*}{\psi}= -1$ occurs at (x_s, y_s) when $\tau_s = \frac{1}{2}\pi\sqrt{y_0}$. The equation for the *switch surface* is:

$$x_s = (\frac{1}{2}\pi + 1)y_s - \frac{1}{2}\pi\sqrt{2y_s}. \tag{12.32}$$

It intersects the y-axis at $(o, \tilde{y}_s)$

$$\tilde{y}_s = \frac{1}{2}\frac{\pi^2 w^2}{(1 + \frac{1}{2}\pi)^2}. \tag{12.33}$$

At all points on (x_s, y_s) on the switch curve:

$$J_x(x_s, y_s) = \frac{1}{\sqrt{y_0} - w} = \frac{1}{\sqrt{2y_s} - w} > 0, \tag{12.34}$$

$$J_y(x_s, y_s) = -\frac{1}{\sqrt{y_0} - w} = -\frac{1}{\sqrt{2y_s} - w}, \tag{12.35}$$

indicating that $\overset{*}{\varphi}_s = \frac{3}{4}\pi$.
For the field of candidate optimal trajectories prior to the switch we have $A(x,y) < 0$ so $\overset{*}{\psi} = -1$ and we have to integrate:

$$\begin{aligned} \dot{J}_x[x(\tau),y(\tau)] &= 0\,, \qquad J_x[x(\tau_s),y(\tau_s)] = \tfrac{1}{\sqrt{2y_s}-w}, \\ \dot{J}_y[x(\tau),y(\tau)] &= \tfrac{\sigma}{2\sqrt{y}}\,, \quad J_y[x(\tau_s),y(\tau_s)] = -\tfrac{1}{\sqrt{2y_s}-w}, \end{aligned} \tag{12.36}$$

and

$$\begin{aligned} \dot{x} &= J_x[x(\tau),y(\tau)]\sqrt{y}/\sigma, \\ \dot{y} &= J_y[x(\tau),y(\tau)]\sqrt{y}/\sigma - w, \end{aligned} \tag{12.37}$$

Now

$$J_x[x(\tau),y(\tau)] = \frac{1}{\sqrt{2y_s}-w} > 0. \tag{12.38}$$

we can obtain $J_y[x(\tau),y(\tau)]$ from ME2:

$$-\sigma\sqrt{y} - wJ_y[x(\tau),y(\tau)] + 1 = 0, \tag{12.39}$$

(since $A(x,y) < 0$ we shall pick the solution with $J_y[x(\tau),y(\tau)] < 0$ for $\tau > \tau_s$).
We can again have an expression for the slope of the candidate optimal trajectories:

$$\frac{dy}{dx} = \pm\frac{\sqrt{(\sqrt{y_o}-w)^2 + w^2 - y}}{\sqrt{y}}. \tag{12.40}$$

We can fortunately[4] substitute

$$y = z^2\,, \quad k = (\sqrt{y_0}-w)^2 + w^2, \tag{12.41}$$

and obtain a closed form solution of equation (12.40)

$$x = \sqrt{y}\sqrt{k-y} - \frac{1}{2}k\sin^{-1}\frac{2y-k}{k} + c. \tag{12.42}$$

Figure 12.1 shows the field of candidate regular optimal trajectories at this stage of the solution.

The safe Contact

Let us first give a heuristic discussion of the possibility of a *safe contact* arc along the y-axis for $y < w^2$.
In the region $y < w^2$ player E dominates $\dot{y}$ and can force it to be negative. Player P can, by choosing $\tilde{\varphi} = \frac{3}{4}\pi$, force the trajectory in this case towards

[4]This was shown to me by my student, Mr. A. Golan.

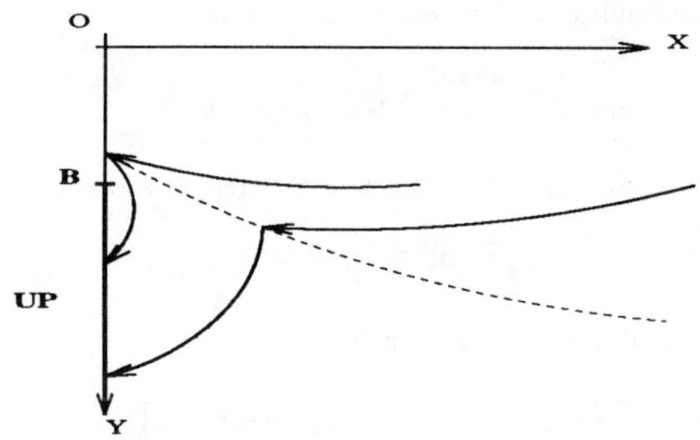

FIGURE 12.1. The first field of regular optimal trajectories

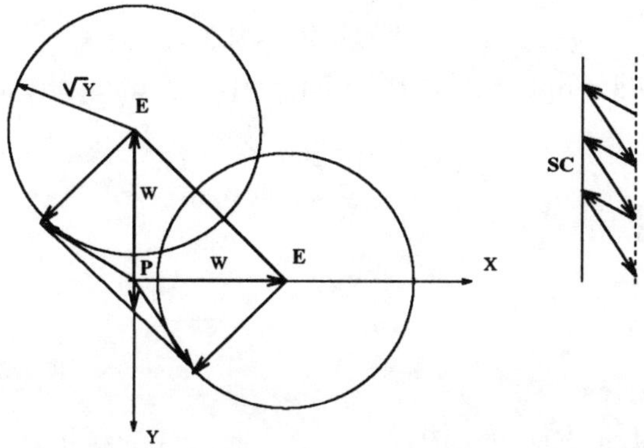

FIGURE 12.2. The chatter equivalent of the safe contact motion

the *NUP*. At the NUP player E will have to change his controls for a duration δ in order to prevent termination. The result is a kind of a *chatter* motion where the segments with velocity $\mathbf{f}(x, y, \frac{3}{4}\pi, -1)$ lead into the NUP and the segments with the velocity $\mathbf{f}(x, y, \frac{3}{4}\pi, +1)$ lead away from it (the corresponding *vectograms* and a typical chatter path are shown in Figure 12.2).

If we let $\delta \to 0$ we may replace the chatter motion by the trajectory that is obtained via the use of the following nonadmissible strategy $\overset{*}{\psi}(\varphi)$:

$$\begin{aligned} \overset{*}{\psi} &= -1\,, && \text{if } \sqrt{y}\cos\varphi > 0, \\ \overset{*}{\psi} &= \tfrac{2\sqrt{y}\cos\varphi}{w} - 1\,, && \text{otherwise.} \end{aligned} \tag{12.43}$$

Notice that all the choices of $\overset{*}{\psi}(\varphi)$ in the domain

$$\frac{3}{2}\pi \geq \varphi \geq \frac{1}{2}\pi, \tag{12.44}$$

yield trajectories that adhere to the y-axis with

$$\dot{y} = \sqrt{y}\sin\varphi - \sqrt{y}\cos\varphi - w, \tag{12.45}$$

We recall Lemma 10.5.2. $\overset{*}{\varphi}$ is determined by:

$$\min_{\frac{3}{2}\pi \geq \varphi \geq \frac{1}{2}\pi} J_y^{(i)}(0,y)[\sqrt{y}\sin\varphi - \sqrt{y}\cos\varphi - w] + 1. \tag{12.46}$$

We notice that for $y > \frac{1}{2}w^2$ player P can guarantee that $\dot{y} > 0$ on the singular arc . so

$$J_y^{(i)}(0,y) < 0,$$

and the candidate optimal controls for player P along the safe contact arc are:

$$(\cos\overset{*}{\varphi}, \sin\overset{*}{\varphi}) \parallel (-1,1), \tag{12.47}$$

or equivalently: $\overset{*}{\varphi} = \frac{3}{4}\pi$.
ME2 provides:

$$J_y^{(i)}(0,y) = -\frac{1}{\sqrt{2y} - w}. \tag{12.48}$$

The Tributaries

The Hamiltonian $H[x, y, \nabla^{(i)}J(x,y), \overset{*}{\varphi}^{(i)}, \psi]$ is linear in ψ. Lemma 10.5.10 says that a transverse junction of the tributaries to the safe contact arc may occur together with the vanishing of the switch function for $\overset{*}{\psi}$ on the tributaries as they reach the safe contact arc.

$$A^{(i)}(0,y) = 0. \tag{12.49}$$

This provides us with the information necessary to evaluate $J_x^{(i)}(0,y)$. Using equation (12.48) we can obtain

$$J_x^{(i)}(0,y) = \frac{1}{\sqrt{2y} - w}, \tag{12.50}$$

and we can compute the candidate optimal tributary trajectories emanating backwards from points $(0, y_0)$ for $\tilde{y}_s > y_0 > w^2/2$ by integrating the equations:

$$\begin{aligned} \dot{J}_x[x(\tau), y(\tau)] &= 0\ , \qquad J_x[x(0), y(0)] = \tfrac{1}{\sqrt{2y_0} - w}, \\ \dot{J}_y[x(\tau), y(\tau)] &= \tfrac{\sigma}{2\sqrt{y}}\ , \quad J_y[x(0), y(0)] = -\tfrac{1}{\sqrt{2y_0} - w}, \end{aligned} \tag{12.51}$$

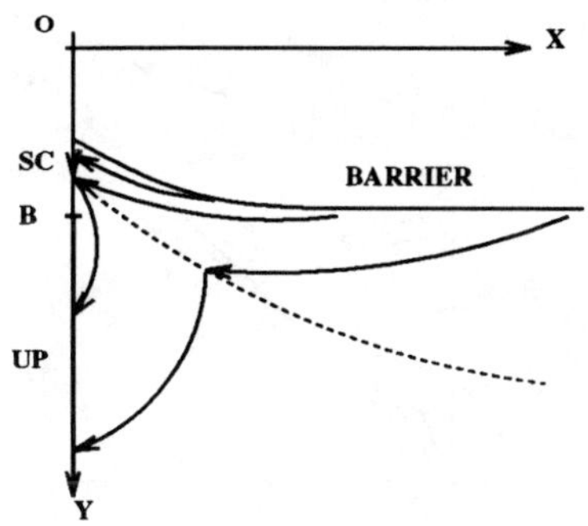

FIGURE 12.3. Optimal trajectories for the Dolichbrachistochrone game

and

$$\begin{aligned} \dot{x} &= J_x[x(\tau), y(\tau)]\sqrt{y}/\sigma \ , \qquad & x(0) = 0, \\ \dot{y} &= J_y[x(\tau), y(\tau)]\sqrt{y}/\sigma - w \ , \qquad & y(0) = y_0. \end{aligned} \tag{12.52}$$

The trajectory that emanates backwards from $(0, \tilde{y}_s)$ is common to the tributary field and to field-1 so the seam between them is smooth!
The tributary that emanates backwards from $(0, \frac{1}{2}w^2)$ is semipermeable (see Problem 8.4.1).
We thus completed to cover the capture-set (bounded by the semipermeable surface and the y-axis) with candidate optimal trajectories that satisfy the necessary conditions (see Figure 12.3).

12.5 Problems

Problem 12.5.1 (The Dolichobrachistochrone problem)

Consider the Dolichobrachistochrone problem. (See Example 12.4.1).

1. Show that $y = w^2$ is a *semipermeable surface* for this system.
2. Examine the permeability of the *corner* at $(0, w^2)$ and show that it leaks.
3. Draw the *vectogram* representation at the corner.
4. Draw the vectogram representation for points on the safe contact arc and draw an equivalent *chatter motion*.

•

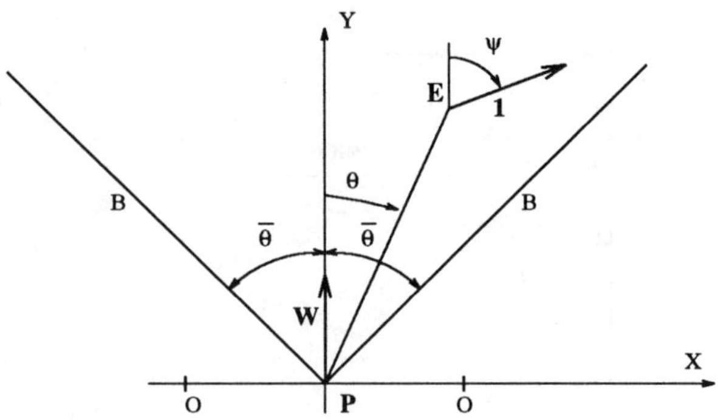

FIGURE 12.4. The Cone Surveillance Evasion problem

Problem 12.5.2 (The Cone Surveillance Evasion game) *[LO79]*

An evader with velocity of magintude 1 and with an unlimlted rate of turn tries to escape from a conic surveillance region in the plane (see Figure 12.4). A pursuer with a speed w and a minimal turn-radius 1 tries to maintain the evader under surveillance as long as possible. (Notice that the dynamics of this game is similar to the *Homicidal Chauffeur game* but the game-set and the objectives of the players are different).

The game-set:
In a Cartesian reference frame:

$$\mathcal{S} \equiv \{(x, y) \mid \frac{y}{|x|} \geq \tan \tilde{\theta}\}.$$

In a polar reference frame:

$$\mathcal{S} \equiv \{(r, \theta) \mid |\theta| \leq \tilde{\theta}\}.$$

The target-set:
In a Cartesian reference frame:

$$\mathcal{C} \equiv \{(x, y) \mid \frac{y}{|x|} = \tan \tilde{\theta}\}.$$

In a polar reference frame:

$$\mathcal{C} \equiv \{(r, \theta) \mid |\theta| = \tilde{\theta}\}.$$

The equations of motion:
In a Cartesian reference frame:

$$\dot{x} = \sin \psi - y w \varphi \;, \qquad |\varphi| \leq 1,$$

$$\dot{y} = \cos \psi + x w \varphi - w.$$

In a polar reference frame:

$$\dot{r} = \cos(\psi - \theta) - w \cos\theta - rw\varphi \ , \quad |\varphi| \le 1,$$

$$r\dot{\theta} = \sin(\psi - \theta) + w \sin\theta .$$

The pursuer chooses φ and the evader chooses ψ.

1. Find the usable part and find the primary optimal trajectories that emanate from it.
2. Investigate the possibility of a *safe contact* along the *NUP* . Who dominates?

•

Problem 12.5.3 (The Obstacle Tag problem)

recall Problem 7.5.6

1. Investigate the safe contact motion of player P backward from the point $r_p = 1$ for the optimal trajectory with $\beta = 1$.
   (a) Find the optimal controls on the safe contact arc.
   (b) Write the adjoint equations for the safe contact arc.
2. Determine the nature of the junction of the tributaries to the safe contact arc.
3. Find equations for the tributaries.

•

Problem 12.5.4

Consider a problem discussed by *P. Bernhard* in [Ber75]. The game takes place in a plane. Two players, P and E, have constant speeds, 1 and $w < 1$. Player P wants to reach a circle of radius 1 about the origin in minimum time subject to the restriction that his distance from player E is at least 1.

A formulation in 4-dimensions

The game-set:

$$\mathcal{S} \equiv \{(\mathbf{r}_p, \mathbf{r}_e) \mid \ |\mathbf{r}_p| \ge 1 \ , \ |\mathbf{r}_p - \mathbf{r}_e| \ge 1 \ \}.$$

The capture-set:

$$\mathcal{C} \equiv \{(\mathbf{r}_p, \mathbf{r}_e) \mid \ |\mathbf{r}_p| = 1 \ , \ |\mathbf{r}_p - \mathbf{r}_e| \ge 1 \ \}.$$

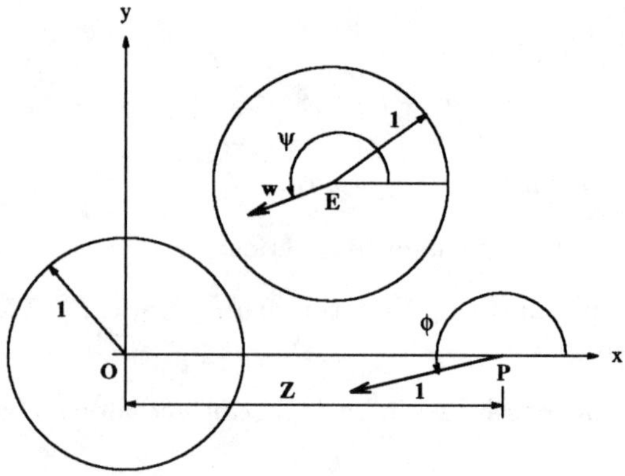

FIGURE 12.5. The reference frame for Example 12.5.4

The equations of motion:

$$\dot{\mathbf{r}}_p = \hat{\beta}_p ,$$

$$\dot{\mathbf{r}}_e = w\hat{\beta}_e .$$

The outcome functional:

$$G(\mathbf{r}_p, \mathbf{r}_e) = 0 \ , \quad L(\mathbf{r}_p, \mathbf{r}_e, \hat{\beta}_p, \hat{\beta}_e) = 1.$$

A formulation in 3-dimensions

Let us have the origin attached to the center of the capture-set and let the x-axis pass through P. (See Fig. 12.5).

The game-set:

$$\mathcal{S} \equiv \{(x, y, z) \mid z \geq 1 \ , \ (x - z)^2 + y^2 \geq 1 \ \}.$$

The capture-set:

$$\mathcal{C} \equiv \{(x, y, z) \mid z = 1 \ , \ (x - z)^2 + y^2 \geq 1 \ \}.$$

The equations of motion:

$$\dot{x} = w \cos \psi + (y \sin \phi)/z,$$

$$\dot{y} = w \sin \psi - (x \sin \phi)/z,$$

$$\dot{z} = \cos \phi.$$

The outcome functional:

$$G(x, y, z) = 0 \ , \quad L(x, y, z, \phi, \psi) = 1.$$

1. Solve for the primary field of candidate optimal trajectories that emanate backwards from the target-set. show that the following quantities
$$\sqrt{J_x^2 + J_y^2},$$
$$\sqrt{J_z^2 + 1/z^2(xJ_y - yJ_x)^2},$$
remain constant along the optimal trajectories.

2. Examine the feasibility of a *safe contact* singular arc along the state constraint surface Q:
$$(x - z)^2 + y^2 = 1.$$
   (a) Identify the curve S on Q where the primary optimal trajectories join the singular arcs.
   (b) determine the nature of the junction between the singular arcs and the regular tributaries.

3. Can we find the *safe contact* controls of the players by assuming a reduced order game on Q in our case?

4. The relation for Γ is (see Bernhard in [Ber77]):
$$\Gamma \frac{\partial Q}{\partial \mathbf{x}} \frac{\partial \overset{*}{\mathbf{f}}}{\partial J_x} \left(\frac{\partial Q}{\partial \mathbf{x}}\right)^T + \overset{*}{\mathbf{f}}^T \frac{\partial^2 Q}{\partial \mathbf{x}^2} \overset{*}{\mathbf{f}} + \frac{\partial Q}{\partial \mathbf{x}} \frac{\partial \overset{*}{\mathbf{f}}}{\partial \mathbf{x}} \overset{*}{\mathbf{f}} - \frac{\partial Q}{\partial \mathbf{x}} \frac{\partial \overset{*}{\mathbf{f}}}{\partial J_x} \frac{\partial \overset{*}{H}}{\partial \mathbf{x}} = 0.$$
Write the specific relation for $\Gamma(t)$ for our case.

•

13

Universal and Focal Surfaces

13.1 Introduction

In the course of conflicts in real life we often have to abandon our original prior plans and time-tables due to unforseen developments. The decision-maker should consider this possibility ahead of time and instruct weather we should stick to the old plan or alternatively adjust it continuously.
A similar phenomenon occur in differential games. Suppose that we deviate from a nominal optimal trajectory that belongs to a field of regular optimal trajectories. In this case it is optimal to follow a neighboring optimal trajectory which is different from the nominal one. A manifold of singular segments of optimal trajectories that lie on a *universal surface* or a *focal surface* represents the other case. If a deviation occurs and the state of the game leaves the singular manifold, it is optimal to follow a trajectory that leads back to the singular surface. In this chapter we shall characterize these surfaces and discuss how to construct them.
Often, when we complete the construction of the primary field of optimal trajectories that emanate from the usable-part, we still have *voids* (regions that are not covered by candidate optimal trajectories), in the game-set (see the examples in chapter 7). Sometimes we can resolve this difficulty by assuming that a *universal surface* (or a *focal surface*) exists in the *void*. In such a case the *void* is coverd by two fields of *tributaries* that emanate from both sides of the singular surface.
Universal surface is the name given by *R. Isaacs* to the case when the tributaries from both sides join transversely.
The case when the tributaries from both sides join tangentially was first discussed by *J. V. Breakwell*. He gave it the name *focal surface*.

13.2 Characterization of US

At a *universal surface* (to be denoted in short as *US*) regular optimal trajectories join *transversely* at both sides (See Figure 9.4).
We can formulate this property as:

$$\begin{aligned} &\mathbf{f}(\mathbf{x}, \overset{*}{\mathbf{u}}^{(1)}, \overset{*}{\mathbf{v}}^{(1)}) \cdot \mathbf{n}(\mathbf{x}) > 0, \\ &\mathbf{f}(\mathbf{x}, \overset{*}{\mathbf{u}}^{(2)}, \overset{*}{\mathbf{v}}^{(2)}) \cdot \mathbf{n}(\mathbf{x}) < 0, \end{aligned} \tag{13.1}$$

where $\mathbf{n(x)}$ is the *normal* to the *universal surface* pointing from side-1 to side-2.
We know from Lemma 10.5.10 (*junction condition 4*) that a transverse junction of the tributaries at their arrival at a singular arc is associated with a *Hamiltonian function* that is linear in component(s) of a control. Let us assume, with no loss of generality, that it is player E who applies controls of the type $\mathbf{v(u)}$ along the *singular arc*. We also assume that

$$\mathbf{v} = \begin{pmatrix} \mathbf{v}_1 \\ \psi \end{pmatrix}.$$

and that the *Hamiltonian function* is linear in the component ψ of the control $\mathbf{v}$ of player E:

$$\mathbf{f(x, u, v)} = \mathbf{a(x, v_1)}\psi + \mathbf{b(x, u, v_1)}. \tag{13.2}$$

We define the case when player E switches ψ but $\overset{*}{\mathbf{v}}{}_1^{(1)} = \overset{*}{\mathbf{v}}{}_1^{(2)}$ as a *regular universal surface*.

Lemma 13.2.1

The function $\nabla J(\mathbf{x})$ is continuous across a *regular universal surface*.

Proof
Let us assume that the lemma is not true and that

$$\nabla J^{(1)}(\mathbf{x}) = \nabla J^{(2)}(\mathbf{x}) + \beta(\mathbf{x})\mathbf{n}(\mathbf{x}). \tag{13.3}$$

with $\beta(\mathbf{x}) > 0$[1].
By Lemma 10.5.10 the switch function for ψ vanishes on the regular optimal trajectories as they arrive at any point $\mathbf{x}$ of the universal surface from *both* sides:

$$\mathbf{J_x^{(i)}(x)} \cdot \mathbf{a}(\mathbf{x}, \overset{*}{\mathbf{v}}_1) = \mathbf{0}.$$

So we have (assuming a *separable Hamiltonian*):

$$H[\mathbf{x}, J_x^{(2)}, \overset{*}{\mathbf{u}}{}^{(2)}, \overset{*}{\mathbf{v}}{}^{(2)}] = H[\mathbf{x}, J_x^{(1)}, \overset{*}{\mathbf{u}}{}^{(1)}, \overset{*}{\mathbf{v}}{}^{(2)}] = 0$$

$$0 = H[\mathbf{x}, J_x^{(1)}, \overset{*}{\mathbf{u}}{}^{(1)}, \overset{*}{\mathbf{v}}{}^{(1)}] = H[\mathbf{x}, J_x^{(2)}, \overset{*}{\mathbf{u}}{}^{(2)}, \overset{*}{\mathbf{v}}{}^{(1)}], \tag{13.4}$$

and:

$$H[\mathbf{x}, J_x^{(2)}, \overset{*}{\mathbf{u}}{}^{(1)}, \overset{*}{\mathbf{v}}{}^{(1)}] = H[\mathbf{x}, J_x^{(2)}, \overset{*}{\mathbf{u}}{}^{(1)}, \overset{*}{\mathbf{v}}{}^{(2)}]. \tag{13.5}$$

---

[1]Lemma 10.4.5 implies that the case where player E utilizes a control law of the type $\mathbf{v(u)}$ along the singular arc is associated with $\beta \geq 0$.

By equation (13.3) we have:

$$\begin{aligned} H[\mathbf{x}, J_x^{(2)}, \overset{*}{\mathbf{u}}^{(1)}, \overset{*}{\mathbf{v}}^{(1)}] = \\ H[\mathbf{x}, J_x^{(1)}, \overset{*}{\mathbf{u}}^{(1)}, \overset{*}{\mathbf{v}}^{(1)}] - \beta(\mathbf{x})\mathbf{f}(\mathbf{x}, \overset{*}{\mathbf{u}}^{(1)}, \overset{*}{\mathbf{v}}^{(1)}) \cdot \mathbf{n}(\mathbf{x}). \end{aligned} \tag{13.6}$$

Recalling that by ME2 in field-1 $H[\mathbf{x}, J_x^{(1)}, \overset{*}{\mathbf{u}}^{(1)}, \overset{*}{\mathbf{v}}^{(1)}] = 0$ and by using relation (13.1) we get that if $\beta(\mathbf{x}) > 0$ then

$$H[\mathbf{x}, J_x^{(2)}, \overset{*}{\mathbf{u}}^{(1)}, \overset{*}{\mathbf{v}}^{(1)}] < 0, \tag{13.7}$$

so by relation (13.5) we have that also:

$$H[\mathbf{x}, J_x^{(2)}, \overset{*}{\mathbf{u}}^{(1)}, \overset{*}{\mathbf{v}}^{(2)}] < 0, \tag{13.8}$$

on the other hand ME1 in field-2 implies that for a deviation from the optimal controls of player P that:

$$H[\mathbf{x}, J_x^{(2)}, \overset{*}{\mathbf{u}}^{(1)}, \overset{*}{\mathbf{v}}^{(2)}] \geq 0. \tag{13.9}$$

•

Corollary 13.2.2

If the minimization problem:

$$\begin{aligned} &\min H[\mathbf{x}, \mathbf{J}_{\mathbf{x}}^{(\mathbf{i})}(\mathbf{x}), \mathbf{u}, \tilde{\mathbf{v}}(\mathbf{u})] \\ &\mathbf{u} \in \{\, \mathbf{u} \mid \mathbf{u} \in \mathcal{U} \;,\; \mathbf{f}[\mathbf{x}, \mathbf{u}, \tilde{\mathbf{v}}(\mathbf{u})] \cdot \mathbf{n}(\mathbf{x}) = 0\}. \end{aligned} \tag{13.10}$$

has a unique solution for both side-1 and side-2 of a *regular universal surface* then:

$$\overset{*}{\mathbf{u}}^{(1)} = \overset{*}{\mathbf{u}}^{(2)} = \overset{*}{\mathbf{u}}\,. \tag{13.11}$$

Proof

By Lemma 10.5.7 we have that

$$\begin{aligned} \overset{*}{\mathbf{u}}^{(1)} &= \tilde{\mathbf{u}}^{(1)}, \\ \overset{*}{\mathbf{u}}^{(2)} &= \tilde{\mathbf{u}}^{(2)}. \end{aligned} \tag{13.12}$$

where $\tilde{\mathbf{u}}^{(i)}$ is the optimal control of player P along the singular arc at side-i. Let us suppose that contrary to the lemma

$$\overset{*}{\mathbf{u}}^{(2)} \neq \overset{*}{\mathbf{u}}^{(1)},$$

or equivalently that

$$\tilde{\mathbf{u}}^{(2)} \neq \tilde{\mathbf{u}}^{(1)}.$$

We have:

$$H[\mathbf{x}, \mathbf{J}_x, \tilde{\mathbf{u}}^{(1)}, \tilde{\mathbf{v}}^{(1)}(\tilde{\mathbf{u}}^{(1)})] = H[\mathbf{x}, \mathbf{J}_x, \tilde{\mathbf{u}}^{(2)}, \tilde{\mathbf{v}}^{(2)}(\tilde{\mathbf{u}}^{(2)})] = 0. \tag{13.13}$$

At a point $\mathbf{x}$ on the singular arc we can consider the pair of controls $[\tilde{\mathbf{u}}^{(1)}, \tilde{\mathbf{v}}^{(2)}(\tilde{\mathbf{u}}^{(1)})]$ as representing a deviation from the optimal control by player P with respect to side-2 so by Lemma 10.5.2

$$H[\mathbf{x}, \mathbf{J}_x, \tilde{\mathbf{u}}^{(1)}, \tilde{\mathbf{v}}^{(2)}(\tilde{\mathbf{u}}^{(1)})] \geq 0, \tag{13.14}$$

and as representing a deviation from the optimal control by player E with respect to side-1 so by Lemma 10.5.3

$$H[\mathbf{x}, \mathbf{J}_x, \tilde{\mathbf{u}}^{(1)}, \tilde{\mathbf{v}}^{(2)}(\tilde{\mathbf{u}}^{(1)})] \leq 0, \tag{13.15}$$

so it follows that:

$$H[\mathbf{x}, \mathbf{J}_x, \tilde{\mathbf{u}}^{(1)}, \tilde{\mathbf{v}}^{(2)}(\tilde{\mathbf{u}}^{(1)})] = 0. \tag{13.16}$$

this means that if $\tilde{\mathbf{u}}^{(2)} \neq \tilde{\mathbf{u}}^{(1)}$ then the problem

$$\begin{array}{l} \min H[\mathbf{x}, \mathbf{J}_{\mathbf{x}}^{(\mathbf{i})}, \mathbf{u}, \tilde{\mathbf{v}}^{(\mathbf{2})}(\mathbf{u})] \\ \mathbf{u} \in \{\, \mathbf{u} \mid \mathbf{u} \in \mathcal{U} \;,\; \mathbf{f}[\mathbf{x}, \mathbf{u}, \tilde{\mathbf{v}}^{(2)}(\mathbf{u})] \cdot \mathbf{n}(\mathbf{x}) = 0\}. \end{array} \tag{13.17}$$

has *nonunique* solution contradicting our assumption.

•

Remarks

1. We can rewrite relations (13.1) that characterize a *regular universal surface* as

$$\begin{array}{c} \mathbf{f}(\mathbf{x}, \overset{*}{\mathbf{u}}, \overset{*}{\mathbf{v}}^{(1)}) \cdot \mathbf{n}(\mathbf{x}) > 0, \\ \\ \mathbf{f}(\mathbf{x}, \overset{*}{\mathbf{u}}, \overset{*}{\mathbf{v}}^{(2)}) \cdot \mathbf{n}(\mathbf{x}) < 0, \end{array} \tag{13.18}$$

   indicating that in spite of the fact that $\beta = 0$ still player E is obliged to use an *information nonadmissible control law* of the type $\mathbf{v} = \mathbf{v}(\mathbf{u})$ in order to satisfy the requirements of Lemma 10.5.

2. We call a *regular universal surface* where player E uses a control of the type $\mathbf{v} = \mathbf{v}(\mathbf{u})$ an *evader's universal surface* and denote it as *EUS*. Similarly, a*regular universal surface* where player P uses a control of the type $\mathbf{u} = \mathbf{u}(\mathbf{v})$ will be called a *pursuer's universal surface* and we shall denote it as *PUS*.

•

The following lemmas, due to *Isaacs*, deal with remarkable properties of the *switch function* $A[\mathbf{x}, \mathbf{J}_x]$ for $\overset{*}{\psi}$ on regular optimal trajectories and at their junction to a *regular universal surface*.

Lemma 13.2.3

On a regular optimal trajectory the function $\overset{\circ}{A}\ [\mathbf{x}(\tau), \mathbf{J}_x(\mathbf{x}(\tau))]$ does not depend on $\overset{*}{\psi}$.

Proof

We shall follow the basic proof given in Isaacs' book.
Recalling Lemma 2.6.1 we shall assume, with no loss in generality, that $L(\mathbf{x}, \mathbf{u}, \mathbf{v}) = 0$.

In the following argument we shall use the following notations abbreviations and conventions for simplifying the algebra of the proof:

1. A *subscript* i will denote *differentiation* with respect to x_i.
   For example:
   $$\frac{\partial \mathbf{J}(\mathbf{x})}{\partial x_i} = \mathbf{J}_i(\mathbf{x}). \tag{13.19}$$
   which will be abbreviated as J_i.

2. A *superscript* k will denote the kth component of a vector.
   For example:
   $$f^k(\mathbf{x}, \mathbf{u}, \mathbf{v}) = a^k(\mathbf{x}, \mathbf{u}, \mathbf{v_1})\psi + b^k(\mathbf{x}, \mathbf{u}, \mathbf{v_1}). \tag{13.20}$$
   which will be abbreviated as $f^k = a^k\psi + b^k$.

3. A repetition of an index within a group of multiplied factors means the summation over each repeated index over its range[2] .
   For example:
   $$A(\mathbf{x}, \nabla J(\mathbf{x})) = \nabla J(\mathbf{x})\cdot \mathbf{a}(\mathbf{x}, \mathbf{u}, \mathbf{v_1}) = \sum_{i=1}^{n} \frac{\partial J(\mathbf{x})}{\partial x_i} a^i(\mathbf{x}, \mathbf{u}, \mathbf{v_1}), \tag{13.21}$$
   which will be abbreviated as $A = J_i a^i$.
   Notice that a repeated index is a *dummy* variable.
   For example:
   $$J_j a^j_i a^i = J_i a^i_j a^j \tag{13.22}$$

---

[2]This is sometimes called *the summation convention*.

•

The adjoint equations on the regular optimal trajectories are:

$$\overset{\circ}{J}_j = J_i[a^i_j \psi + b^i_j], \tag{13.23}$$

from which we can obtain that:

$$\overset{\circ}{J}_j \; a^j = J_i a^i_j a^j \psi + J_i b^i_j a^j . \tag{13.24}$$

Differentiating A with respect to τ we get:

$$\overset{\circ}{A} = \overset{\circ}{J}_j \; a^j - J_j a^j_i f^i = \overset{\circ}{J}_j \; a^j - J_j a^j_i a^i \psi - J_j a^j_i b^i . \tag{13.25}$$

Substituting from equation (13.24) into equation (13.25) and recalling that a repeated index is a *dummy* variable we get

$$\overset{\circ}{A} = J_i[b^i_j a^j - a^i_j b^j]. \tag{13.26}$$

which proves that $\overset{\circ}{A}\,[\mathbf{x}(\tau), \mathbf{J}_x(\mathbf{x}(\tau))]$ does not depend on $\overset{*}{\psi}$.

•

Lemma 13.2.4

It is necessary that $\overset{\circ}{A}\,(\mathbf{x}, \mathbf{J}_x) = 0$ at points where tributaries join singular arcs on a *regular universal surface*.

Proof

$\nabla J(\mathbf{x})$ is continuous at the junction (see Lemma 13.2.1).
If the assertion of the lemma is not true it follows that at a point $\mathbf{x}$ on the universal surface, for sufficiently small τ , the *switch function* $A(\tau)$ might have the same sign at the tributaries that emanate to both sides of the *universal surface*. If this is the case we shall have that the velocity vectors of the tributaries of side-1 and side-2 are equal at the junction point. This contradicts equation (13.18). which state that the tributaries join a *universal surface* transversely.

•

If we denote

$$c^i = [b^i_j a^j - a^i_j b^j], \tag{13.27}$$

we could summarize the necessary conditions at the junction to the *US* by requiring that:

$$\begin{aligned} a^i J_i &= 0, \qquad \text{representing } A(\mathbf{x}, \mathbf{J}_x) = 0, \\ b^i J_i &= 0, \qquad \text{representing } H[\mathbf{x}, \mathbf{J}_x, \overset{*}{\mathbf{u}}, \overset{*}{\mathbf{v}}^{(i)}] = 0, \\ c^i J_i &= 0, \qquad \text{representing } \overset{\circ}{A}\,(\mathbf{x}, \mathbf{J}_x) = 0, \end{aligned} \tag{13.28}$$

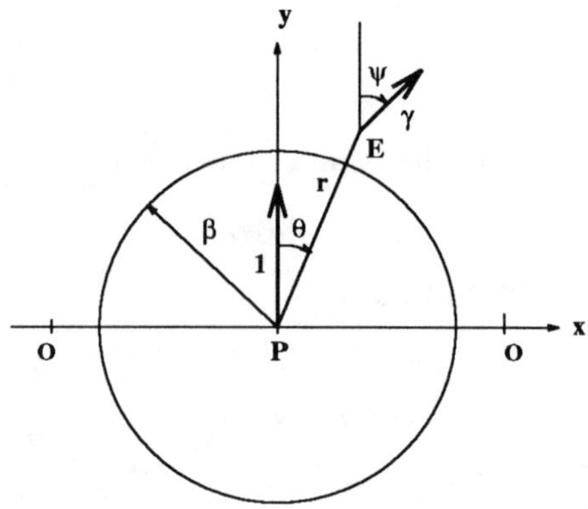

FIGURE 13.1. The reference frames for the Homicidal Chauffeur problem

13.3 Examples

Example 13.3.1 (The Homicidal Chauffeur) *[Isa75,Mer71]*

In an infinite planar *parking lot* a circular *car* P strives to knock a slower *pedestrian* E. The *car* has a radius β, a maximal speed of magnitude 1 and a minimal turn radius of magnitude 1. The *pedestrian* has a maximal speed of magnitude $\gamma < 1$ and a capability to perform instantaneous turns. *R. Isaacs* invented this problem in his efforts to analyze nontrivial pursuit evasion real life problems, (such as *air-combat*), where one side has a speed advantage while the other is more maneuverable.
Let us formulate the game in a *relative frame* where the origin is attached to player P and the y-axis points in the direction of the velocity of player P (see Figure 13.1).

The game-set:

$$\mathcal{S} \equiv \{(x, y) \mid x^2 + y^2 \geq \beta^2\} \quad \text{in a Cartesian frame.}$$

$$\mathcal{S} \equiv \{(r, \theta) \mid r \geq \beta\} \quad \text{in a polar frame.}$$

The target-set:

$$\mathcal{S} \equiv \{(x, y) \mid x^2 + y^2 = \beta^2\} \quad \text{in a Cartesian frame.}$$

$$\mathcal{S} \equiv \{(r, \theta) \mid r = \beta\} \quad \text{in a polar frame.}$$

The equations of motion:
in a Cartesian frame:

$$\begin{aligned} \dot{x} &= \gamma \sin \psi - y\varphi, \qquad |\varphi| \leq 1, \\ \dot{y} &= \gamma \cos \psi - 1 + x\varphi, \end{aligned} \tag{13.29}$$

and in a polar frame:

$$\dot{r} = \gamma\cos(\psi - \theta) - \cos\theta,$$
$$r\dot{\theta} = \gamma\sin(\psi - \theta) + \sin\theta - \varphi, \qquad |\varphi| \le 1. \tag{13.30}$$

$1 > \gamma > 0$ is a parameter. Player P chooses φ and player E chooses ψ.

The Outcome Functional:

$$G(x, y) = 0\ , \qquad L(x, y, \varphi, \psi) = 1.$$

In *pursuit evasion* games it is the pursuer, player P, who prefers to terminate and to minimize the time-to-capture. Player E prefers nontermination over any numerical outcome.

remark

1. In this formulation we assumed that the magnitude of the minimal radius of turn of the pursuer and the magnitude of the maximal speed of the evader are equal to unity. Indeed this is a normalized version and the two parameters of the game γ and β can be considered as the ratios between the speeds of the pursuer and the evader and the ratio between the radius of the surveillance circle and the minimal radius of turn respectively.

The partitioning of the target-set

The normal to the target-set pointing into the game-set is $\mathbf{n}(\beta,\theta) = \hat{\mathbf{e}}_r$ (where $\hat{\mathbf{e}}_r$ is a unit vector in the r-direction).

We have:

$$\max_{\psi}\,[\gamma\cos(\psi - \theta) - \cos\theta] = \gamma - \cos\theta.$$

So:

$$\mathrm{UP} \equiv \{(\beta,\theta) \mid |\theta| < \cos^{-1}\gamma\}.$$

$$\mathrm{NUP} \equiv \{(\beta,\theta) \mid |\theta| > \cos^{-1}\gamma\}.$$

$$\mathrm{BUP} \equiv \{(\beta,\theta) \mid |\theta| = \cos^{-1}\gamma\}.$$

Candidate optimal control laws

The equations of motion are *separable* so we can express the Isaacs equation ME1 as:

$$\min_{\varphi}\max_{\psi}[(\gamma\sin\psi - y\varphi)J_x + (\gamma\cos\psi - 1 + x\varphi)J_y + 1].$$

or:

$$\min_{\varphi}\max_{\psi}\{[\gamma\cos(\psi-\theta) - \cos\theta]J_r + \tfrac{1}{r}[\gamma\sin(\psi-\theta) + \sin\theta - \varphi]J_\theta + 1\}.$$

Let $A_c(x,y)$ be the *switch function* for $\overset{*}{\varphi}$ in the Cartesian representation.

$$A_c(x,y) = xJ_y(x,y) - yJ_x(x,y).$$

The candidate optimal control laws will be:

$$(\cos\overset{*}{\psi}, \sin\overset{*}{\psi}) \parallel (J_x, J_y),$$

$$\overset{*}{\varphi} = \text{sign} A_c(x,y)\ , \qquad \text{for} \ \ A_c(x,y) \neq 0, \tag{13.31}$$

$$\overset{*}{\varphi} = \text{any admissible} \ \ \varphi\ , \qquad \text{for} \ \ A_c(x,y) = 0.$$

Similarly for the representation in polar coordinates we have:

$$A_p(r,\theta) = J_\theta(r,\theta),$$

$$(\cos(\overset{*}{\psi} - \theta), (\sin(\overset{*}{\psi} - \theta)) \parallel (J_r, J_\theta),$$

$$\overset{*}{\varphi} = \ \text{sign} A_p(r,\theta)\ , \qquad \text{for} \ \ A_p(r,\theta) \neq 0, \tag{13.32}$$

$$\overset{*}{\varphi} = \text{any admissible} \ \ \varphi\ , \qquad \text{for} \ \ A_p(r,\theta) = 0.$$

Notice that the switch in $\overset{*}{\varphi}$ takes place when $\nabla J(x,y)$ passes through P.

The Adjoint Equations

$$\overset{\circ}{J}_x\ [x(\tau), y(\tau)] = J_y[x(\tau), y(\tau)]\ \overset{*}{\varphi},$$
$$\overset{\circ}{J}_y\ [x(\tau), y(\tau)] = -J_x[x(\tau), y(\tau)]\ \overset{*}{\varphi}, \tag{13.33}$$

and this provides:

$$J_x[x(\tau), y(\tau)]\ \overset{\circ}{J}_x\ [x(\tau), y(\tau)] + J_y[x(\tau), y(\tau)]\ \overset{\circ}{J}_y\ [x(\tau), y(\tau)] = 0. \tag{13.34}$$

which means that $\nabla J(x,y)$ is constant along regular segments of optimal trajectories. We also have:

$$\overset{\circ}{A}\ (x,y) = -J_x(x,y). \tag{13.35}$$

The tangent $\mathbf{t}(\beta,\theta)$ to the target-set at (β,θ) is $\mathbf{t} = \hat{\mathbf{e}}_\theta$ in the polar representation and $\mathbf{t} = y\hat{\mathbf{e}}_x - x\hat{\mathbf{e}}_y$ in the Cartesian frame.
We have that $\nabla G(r,\theta) = 0$ so:

$$J_\theta(\beta,\theta) = 0.$$

From ME-2 we obtain in the polar case:

$$\gamma|J_r(\beta,\theta)| - J_r(\beta,\theta)\cos\theta + 1 = 0.$$

The requirement that player E wants to avoid termination at the points of the UP requires that $\cos(\overset{*}{\psi} - \theta) \geq 0$ so that $J_r(\beta,\theta) \geq 0$.
We obtain:

$$J_r(\beta,\theta) = -(\gamma - \cos\theta)^{-1}\ , \quad \text{for} \quad |\theta| < \cos^{-1}\gamma, \tag{13.36}$$

and in the Cartesian reference frame (here θ_f is a parameter):

$$J_x(\beta\sin\theta_f, \beta\cos\theta_f) = \frac{\sin\theta_f}{(\cos\theta_f - \gamma)}\ , \text{ for } |\theta_f| < \cos^{-1}\gamma, \tag{13.37}$$

$$J_y(\beta\sin\theta_f, \beta\cos\theta_f) = \frac{\cos\theta_f}{(\cos\theta_f - \gamma)}\ , \text{ for } |\theta_f| < \cos^{-1}\gamma. \tag{13.38}$$

Candidate optimal trajectories
Integrating the adjoint equations we get:

$$J_x[x(\tau), y(\tau)] = \frac{\sin(\theta_f + \varphi\tau)}{(\cos\theta_f - \gamma)}, \tag{13.39}$$

$$J_y[x(\tau), y(\tau)] = \frac{\cos(\theta_f + \varphi\tau)}{(\cos\theta_f - \gamma)}. \tag{13.40}$$

The equations of the candidate optimal trajectories are:

$$\begin{aligned} \dot{x} &= \gamma\sin(\theta_f + \overset{*}{\varphi}\,\tau) + y\,\overset{*}{\varphi}, \\ \dot{y} &= \gamma\cos(\theta_f + \overset{*}{\varphi}\,\tau) - x\,\overset{*}{\varphi} + 1. \end{aligned} \tag{13.41}$$

The candidate optimal trajectories are given by:

$$\begin{aligned} x(\tau) &= \overset{*}{\varphi}(1 - \cos\tau) + \beta\sin\theta_f\cos\tau + \overset{*}{\varphi}\,\beta\cos\theta_f\sin\tau - \gamma\tau\sin(\theta_f + \overset{*}{\varphi}\,\tau), \\ y(\tau) &= (1 - \overset{*}{\varphi}\,\beta\sin\overset{*}{\varphi}\,f)\sin\tau + \beta\cos\theta_f\cos\tau - \gamma\tau\cos(\theta_f + \overset{*}{\varphi}\,\tau). \end{aligned} \tag{13.42}$$

In the first quadrant, where $\cos^{-1}\gamma > \theta_f > 0$, we have that $\overset{*}{\varphi} = +1$
The candidate optimal trajectories for $\overset{*}{\varphi} = +1$ are:

$$\begin{aligned} x(\tau) &= (1 - \cos\tau) + (\beta - \gamma\tau)\sin(\theta_f + \tau), \\ y(\tau) &= \sin\tau + (\beta - \gamma\tau)\cos(\theta_f + \tau). \end{aligned} \tag{13.43}$$

In the fourth quadrant, where $\cos^{-1}\gamma > -\theta_f > 0$, we have that $\overset{*}{\varphi} = -1$
The candidate optimal trajectories for $\overset{*}{\varphi} = -1$ are:

$$\begin{aligned} x(\tau) &= -(1 - \cos\tau) + (\beta - \gamma\tau)\sin(\theta_f - \tau), \\ y(\tau) &= \sin\tau + (\beta - \gamma\tau)\cos(\theta_f - \tau). \end{aligned} \tag{13.44}$$

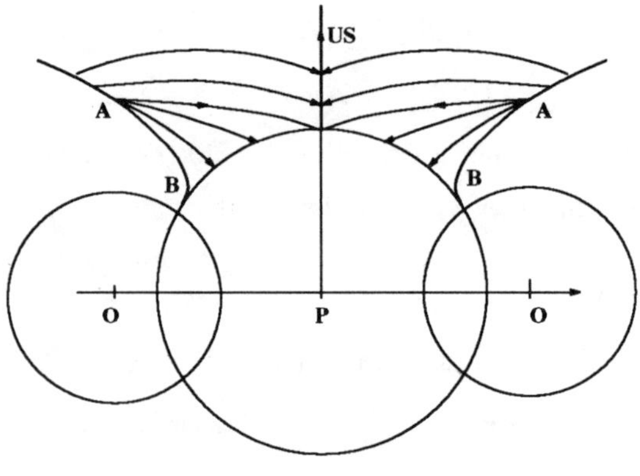

FIGURE 13.2. First field of candidate optimal trajectories

A pictorial description of the field of candidate optimal trajectories that we obtained thus far is given in Figure 13.2 (for the case $\beta = 0.8,\ \gamma = 0.5$). We find that the trajectories that emanate from the BUP points B are candidate *barriers* and that all the trajectories that emanate from other points on the UP pass through the points A on the *barriers*.

We realize that we have a *void* and that the two extreme trajectories arrive at the point $(0,\beta)$ with $A(0,\beta) = 0$ and $\overset{\circ}{A}(0,\beta) = 0$.

We can try to construct a *regular universal surface* beginning at the point $(0,\beta)$.

Recalling relations (13.28) we have:

$$
\begin{aligned}
-yJ_x(x,y) \qquad & +xJ_y(x,y) & = 0, \\
\gamma \sin \overset{*}{\psi}\ J_x(x,y) \quad & +(\gamma \cos \overset{*}{\psi} -1)J_y(x,y) & = -1, \\
(1-\gamma \cos \overset{*}{\psi})J_x(x,y) \quad & +\gamma \sin \overset{*}{\psi}\ J_y(x,y) & = 0.
\end{aligned}
\tag{13.45}
$$

It is easy to see that on the singular arc that start at the point $(0,\beta)$ we must have:

$$
\overset{*}{\varphi} = 0\ , \qquad \overset{*}{\psi} = 0, \tag{13.46}
$$

so that the y-axis constitutes a US for $y \geq \beta$.

We also find that $J_y(0,y) = \frac{1}{(\gamma-1)}$ and $J_x(0,y) = 0$ on the US. We can obtain the tributaries to the US by integrating the following system of equations backwards from points on the *US*:

$$
\begin{aligned}
&\overset{\circ}{x} = \gamma \sin \overset{*}{\psi} - y \overset{*}{\varphi}, && x(0) = 0,\\
&\overset{\circ}{y} = \gamma \cos \overset{*}{\psi} + x \overset{*}{\varphi} - 1, && y(0) = y_0,\\
&\overset{\circ}{J_x} = - \overset{*}{\varphi} J_y, && J_x(x(0), y(0)) = 0,\\
&\overset{\circ}{J_y} = \overset{*}{\varphi} J_x, && J_y(x(0), y(0)) = (\gamma - 1)^{-1}.
\end{aligned}
\tag{13.47}
$$

The field of tributaries in the first quadrant is given by:

$$
\begin{aligned}
x(\tau) &= 1 - \cos \tau + (y_0 - \gamma\tau) \sin \tau,\\
y(\tau) &= \sin \tau + (y_0 - \gamma\tau) \cos \tau,\\
J_x(\tau) &= (1 - \gamma)^{-1} \sin \tau,\\
J_y(\tau) &= (1 - \gamma)^{-1} \cos \tau.
\end{aligned}
\tag{13.48}
$$

We observe that the tributary field and the primary field of candidate optimal trajectories join smoothly.

•

13.3.1 The Chatter Equivalent

Let us examine what happens in our example if on the *universal surface* player P is not allowed to utilize an *nonadmissible control law* of the type $\varphi = \varphi(\psi)$.
In the first quadrant, (denoted as *region 1*), we have:

$$0 < \theta, \quad \overset{*}{\varphi}{}^{(1)} = 1, \quad \sin \overset{*}{\psi}{}^{(1)} > 0,$$

and in the second Quadrant, (denoted as *region 2*), we have:

$$\theta < 0, \quad \overset{*}{\varphi}{}^{(2)} = -1, \quad \sin \overset{*}{\psi}{}^{(2)} < 0.$$

At the point O_1 on the *US* player P, who does not have an admissible strategy to prevent the crossing of the *US*, might play $\overset{*}{\varphi}{}^{(1)}$ in an *open-loop* fashion for a duration δ.
If player E plays $\overset{*}{\psi}{}^{(2)}$ during that period we get:

$$\dot{\theta}(r, 0, \overset{*}{\varphi}{}^{(1)}, \overset{*}{\psi}{}^{(2)}) = \gamma \sin \overset{*}{\psi}{}^{(2)} - 1 < 0,$$

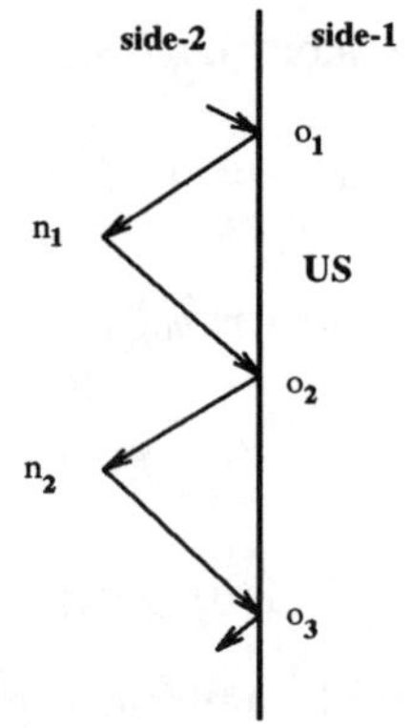

FIGURE 13.3. The chatter equivalent for the motion on the Universal Surface

and the trajectory leads to region-2 where the control $\overset{*}{\varphi}^{(1)}$ may be nonoptimal for player P.
When the duration δ elapses and the trajectory reached the point n_1 inside region-2 (see Figure 13.3), player P resumes playing optimally in a *closed loop* fashion and by playing $\overset{*}{\varphi}^{(2)}$ the trajectory leads back to the *US* and joins it transversely at the point o_2.
Here player P faces the same difficulties and may play open loop for another duration δ and so forth.
The result is a kind of a *chatter* where the segments generated by the pair $(\overset{*}{\varphi}^{(2)}, \overset{*}{\psi}^{(2)})$ lead towards the *US* and the segments generated by the pair $(\overset{*}{\varphi}^{(1)}, \overset{*}{\psi}^{(2)})$ lead away from it. Both players apply *admissible control laws.*
If we let $\delta \to 0$ the *chatter motion* converges to the singular arc on the *US* and the *chatter velocity* (the mix between the velocities on the pair of segments) converges to the velocity on the singular arc generated by the *nonadmissible control law* $\varphi = \varphi(\psi)$

•

13.4 Characterization of Focal Surfaces

At a *focal surface* (to be denoted as *FS*) regular optimal trajectories arrive and join tangentially from both sides. We can formulate this property as:

$$\begin{aligned}\mathbf{f}(\mathbf{x}, \overset{*}{\mathbf{u}}^{(1)}, \overset{*}{\mathbf{v}}^{(1)}) \cdot \mathbf{n}(\mathbf{x}) &= 0, \\ \mathbf{f}(\mathbf{x}, \overset{*}{\mathbf{u}}^{(2)}, \overset{*}{\mathbf{v}}^{(2)}) \cdot \mathbf{n}(\mathbf{x}) &= 0.\end{aligned} \tag{13.49}$$

where $\mathbf{n}(\mathbf{x})$ is the *normal* to the *focal surface* pointing from side-1 to side-2.

Lemma 13.4.1

1. $\nabla J(\mathbf{x})$ is discontinuous across a *focal surface*.
2. The optimal controls of the players at each side of the FS are different.

$$\overset{*}{\mathbf{u}}^{(1)} \neq \overset{*}{\mathbf{u}}^{(2)} , \quad \overset{*}{\mathbf{v}}^{(1)} \neq \overset{*}{\mathbf{v}}^{(2)} , \tag{13.50}$$

Proof

Let us assume that the lemma is not true and that $\beta = 0$:

$$\nabla J^{(1)}(\mathbf{x}) = \nabla J^{(2)}(\mathbf{x}) = \nabla J(\mathbf{x}). \tag{13.51}$$

From ME1 at side-1 and side-2 we can deduce that:

$$H[\mathbf{x}, \mathbf{J}_x, \overset{*}{\mathbf{u}}^{(1)}, \overset{*}{\mathbf{v}}^{(1)}] = H[\mathbf{x}, \mathbf{J}_x, \overset{*}{\mathbf{u}}^{(2)}, \overset{*}{\mathbf{v}}^{(2)}] = 0. \tag{13.52}$$

Considering $\overset{*}{\mathbf{u}}^{(1)}$ as a deviation from the optimal controls at side-2 by player P we get from ME1:

$$H[\mathbf{x}, \mathbf{J}_x, \overset{*}{\mathbf{u}}^{(1)}, \overset{*}{\mathbf{v}}^{(2)}] \geq 0, \tag{13.53}$$

and considering $\overset{*}{\mathbf{v}}^{(2)}$ as a deviation from the optimal controls at side-1 by player E we get from ME1:

$$H[\mathbf{x}, \mathbf{J}_x, \overset{*}{\mathbf{u}}^{(1)}, \overset{*}{\mathbf{v}}^{(2)}] \leq 0, \tag{13.54}$$

so

$$H[\mathbf{x}, \mathbf{J}x, \overset{*}{\mathbf{u}}^{(1)}, \overset{*}{\mathbf{v}}^{(2)}] = 0, \tag{13.55}$$

and similarly

$$H[\mathbf{x}, \mathbf{J}_x, \overset{*}{\mathbf{u}}^{(2)}, \overset{*}{\mathbf{v}}^{(1)}] = 0. \tag{13.56}$$

It is easy to see that if $\overset{*}{\mathbf{u}}^{(1)} \neq \overset{*}{\mathbf{u}}^{(2)}$ and/or $\overset{*}{\mathbf{v}}^{(1)} \neq \overset{*}{\mathbf{v}}^{(2)}$ one of the problems

$$\max_{\mathbf{v} \in \mathcal{V}} H[\mathbf{x}, \mathbf{J}_x, \overset{*}{\mathbf{u}}, \mathbf{v}] , \qquad \min_{\mathbf{u} \in \mathcal{U}} H[\mathbf{x}, \mathbf{J}_x, \mathbf{u}, \overset{*}{\mathbf{v}}], \tag{13.57}$$

will have *nonunique* solutions. Recall that due to the convexity of the *hodograph* sets those problems should have a unique solution unless the Hamiltonian is linear in component(s) of a control vector and the corresponding switch function vanishes. This completes the proof of the first part of the lemma.

Suppose now that:

$$\overset{*}{\mathbf{u}}^{(1)} \neq \overset{*}{\mathbf{u}}^{(2)} , \quad \overset{*}{\mathbf{v}}^{(1)} = \overset{*}{\mathbf{v}}^{(2)} = \overset{*}{\mathbf{v}} .$$

The tangential junction of the tributaries at side 1 provides:

$$\mathbf{f}(\mathbf{x}, \overset{*}{\mathbf{u}}^{(1)}, \overset{*}{\mathbf{v}}) \cdot \mathbf{n}(\mathbf{x}) = 0. \tag{13.58}$$

Using equation (10.12) and applying ME1 at side-1 we get:

$$H(\mathbf{x}, \mathbf{J}_x^{(2)}, \overset{*}{\mathbf{u}}^{(1)}, \overset{*}{\mathbf{v}}) = H(\mathbf{x}, \mathbf{J}_x^{(1)}, \overset{*}{\mathbf{u}}^{(1)}, \overset{*}{\mathbf{v}}) = 0, \tag{13.59}$$

so both $\overset{*}{\mathbf{u}}^{(1)}$ and $\overset{*}{\mathbf{u}}^{(2)}$ are *different* solutions to the problem

$$\min_{\mathbf{u} \in \mathcal{U}} H[\mathbf{x}, \mathbf{J}_x^{(2)}, \mathbf{u}, \overset{*}{\mathbf{v}}], \tag{13.60}$$

contrary to our assumptions.

●

Lemma 13.4.2

At a *focal surface*

$$\begin{aligned} \mathbf{f}(\mathbf{x}, \overset{*}{\mathbf{u}}^{(1)}, \overset{*}{\mathbf{v}}^{(2)}) \cdot \mathbf{n}(\mathbf{x}) \neq 0, \\ \mathbf{f}(\mathbf{x}, \overset{*}{\mathbf{u}}^{(2)}, \overset{*}{\mathbf{v}}^{(1)}) \cdot \mathbf{n}(\mathbf{x}) \neq 0. \end{aligned} \tag{13.61}$$

Proof

Considering $\overset{*}{\mathbf{u}}^{(1)}$ as a deviation from the optimal controls at side-2 by player P we get from ME1:

$$H[\mathbf{x}, J_x^{(2)}, \overset{*}{\mathbf{u}}^{(1)}, \overset{*}{\mathbf{v}}^{(2)}] \geq 0, \tag{13.62}$$

and considering $\overset{*}{\mathbf{v}}^{(2)}$ as a deviation from the optimal controls at side-1 by player E we get from ME1:

$$H[\mathbf{x}, J_x^{(1)}, \overset{*}{\mathbf{u}}^{(1)}, \overset{*}{\mathbf{v}}^{(2)}] \leq 0. \tag{13.63}$$

If the lemma is not true we can assume (with no loss of generality) that:

$$\mathbf{f}(\mathbf{x}, \overset{*}{\mathbf{u}}^{(1)}, \overset{*}{\mathbf{v}}^{(2)}) \cdot \mathbf{n}(\mathbf{x}) = 0. \tag{13.64}$$

So by using equation (10.12) we get

$$H[\mathbf{x}, J_x^{(2)}, \overset{*}{\mathbf{u}}^{(1)}, \overset{*}{\mathbf{v}}^{(2)}] = H[\mathbf{x}, J_x^{(1)}, \overset{*}{\mathbf{u}}^{(1)}, \overset{*}{\mathbf{v}}^{(2)}] = 0, \tag{13.65}$$

implying that both $\overset{*}{\mathbf{u}}^{(1)}$ and $\overset{*}{\mathbf{u}}^{(2)}$ are *different* solutions to the problem

$$\min_{\mathbf{u} \in \mathcal{U}} H[\mathbf{x}, \mathbf{J}_x^{(2)}, \mathbf{u}, \overset{*}{\mathbf{v}}^{(2)}], \tag{13.66}$$

and both $\overset{*}{\mathbf{v}}^{(1)}$ and $\overset{*}{\mathbf{v}}^{(2)}$ are *different* solutions to the problem

$$\max_{\mathbf{v} \in \mathcal{V}} H[\mathbf{x}, \mathbf{J}_x^{(1)}, \overset{*}{\mathbf{u}}^{(1)}, \mathbf{v}], \tag{13.67}$$

contrary to our assumptions.

•

When $\beta > 0$ we have, by lemma 10.4.5, that

$$\begin{aligned} \mathbf{f}(\mathbf{x}, \overset{*}{\mathbf{u}}^{(1)}, \overset{*}{\mathbf{v}}^{(2)}) \cdot \mathbf{n}(\mathbf{x}) &< 0 \\ \mathbf{f}(\mathbf{x}, \overset{*}{\mathbf{u}}^{(2)}, \overset{*}{\mathbf{v}}^{(1)}) \cdot \mathbf{n}(\mathbf{x}) &> 0 \end{aligned} \tag{13.68}$$

In this case both the controls $\overset{*}{\mathbf{v}}^{(1)}$ and $\overset{*}{\mathbf{v}}^{(2)}$ of player E *fail* to satisfy the requirements of Lemma 10.5. Player E has to apply a *nonadmissible control law* of the type $\mathbf{v} = \mathbf{v}(\mathbf{u})$ in order to realize the motion along the singular arc and satisfy the conditions of Lemma 10.5 for optimality. (This is the case of an *evader's focal surface* denoted as *EFS*).
Similarly, if $\beta < 0$

$$\begin{aligned} \mathbf{f}(\mathbf{x}, \overset{*}{\mathbf{u}}^{(1)}, \overset{*}{\mathbf{v}}^{(2)}) \cdot \mathbf{n}(\mathbf{x}) &> 0, \\ \mathbf{f}(\mathbf{x}, \overset{*}{\mathbf{u}}^{(2)}, \overset{*}{\mathbf{v}}^{(1)}) \cdot \mathbf{n}(\mathbf{x}) &< 0, \end{aligned} \tag{13.69}$$

and it is the case where player P will have to apply the *information non-admissible control law* of the type $\mathbf{u} = \mathbf{u}(\mathbf{v})$ in order to realize an optimal motion along the singular arc. (This is the case of an *pursuer's focal surface* denoted as *PFS*).

13.5 Construction of Focal Surfaces

We try a construction of a *focal surface*, if two manifolds of regular optimal trajectories (that belong to an interim solution) arrive at some (n-2)-dimensional surface D from both sides tangentially with a jump discontinuity in $\nabla J_x(\mathbf{x})$. Unlike the cases of *safe contact*, the location of the *FS* that may emanate from D is unknown ahead of time and has to be found as part of the construction of the solution. We have to use the adjoint equations for singular arcs[3], equation (10.55). (See *P. Bernhard* in [Ber77]).
Using the notations of Lemma 10.6.3 we have for *focal surfaces* that along

[3] In 2-dimensional games we can often use ME2 for the adjacent region in order to find the adjoint vector at the junction with the tributaries.

a singular arc adjacent to side i (where $i = 1, 2$):

$$\dot{\mathbf{J}}_x^{(i)} = -\frac{\partial \overset{*}{H}^{(i)}}{\partial x} + \Gamma^{(i)}(t)(\mathbf{J}_x^{(1)} - \mathbf{J}_x^{(2)}). \tag{13.70}$$

We also have the requirement that the tributaries arrive tangent to the *focal surface* from both sides:

$$\begin{aligned} (\mathbf{J}_x^{(1)} - \mathbf{J}_x^{(2)}) \cdot \overset{*(1)}{\mathbf{f}} &= 0, \\ (\mathbf{J}_x^{(1)} - \mathbf{J}_x^{(2)}) \cdot \overset{*(2)}{\mathbf{f}} &= 0. \end{aligned} \tag{13.71}$$

We first notice that:

$$\frac{d}{dt} \overset{*(i)}{\mathbf{f}} = \overset{*(i)}{\mathbf{f}_x} \overset{*(i)}{\mathbf{f}} + \overset{*(i)}{\mathbf{f}_{J_x}} \dot{\mathbf{J}}_x^{(i)}. \tag{13.72}$$

Next we differentiate relations (13.71) and substitute from equation (13.72) and (13.70): to get the following system of equations from which we can compute $\Gamma^{(1)}(\tau)$ and $\Gamma^{(2)}(\tau)$:

$$\begin{aligned} (\overset{*(1)}{\mathbf{f}})^T (\mathbf{J}_{xx}^{(1)} - \mathbf{J}_{xx}^{(2)}) \overset{*(1)}{\mathbf{f}} + {} & (\mathbf{J}_x^{(1)} - \mathbf{J}_x^{(2)}) \overset{*(1)}{\mathbf{f}_x} \overset{*(1)}{\mathbf{f}} - (\mathbf{J}_x^{(1)} - \mathbf{J}_x^{(2)}) \overset{*(1)}{\mathbf{f}_{J_x}} \overset{*}{\mathbf{H}}{}_x^{(1)} \\ & + \Gamma^{(1)} (\mathbf{J}_x^{(1)} - \mathbf{J}_x^{(2)}) \overset{*(1)}{\mathbf{f}_{J_x^{(1)}}} (\mathbf{J}_{xx}^{(1)} - \mathbf{J}_{xx}^{(2)})^T = 0, \\ (\overset{*(2)}{\mathbf{f}})^T (\mathbf{J}_{xx}^{(1)} - \mathbf{J}_{xx}^{(2)}) \overset{*(2)}{\mathbf{f}} + {} & (\mathbf{J}_x^{(1)} - \mathbf{J}_x^{(2)}) \overset{*(2)}{\mathbf{f}_x} \overset{*(2)}{\mathbf{f}} - (\mathbf{J}_x^{(1)} - \mathbf{J}_x^{(2)}) \overset{*(2)}{\mathbf{f}_{J_x}} \overset{*}{\mathbf{H}}{}_x^{(2)} \\ & + \Gamma^{(2)} (\mathbf{J}_x^{(1)} - \mathbf{J}_x^{(2)}) \overset{*(2)}{\mathbf{f}_{J_x^{(2)}}} (\mathbf{J}_{xx}^{(1)} - \mathbf{J}_{xx}^{(2)})^T = 0. \end{aligned} \tag{13.73}$$

13.6 An Example of a Focal Surface

Example 13.6.1 (The Circular Wall Pursuit)

Let us consider the *Circular Wall Pursuit* problem. It is a problem of pursuit evasion that takes place inside a circle with a radius of magnitude 1.

The evader, player E, is confined to move on the circumference of the circle. His velocity has the magnitude of 1.

The pursuer, player P, moves in the circle with a speed of magnitude $w > 1$ and is capable to turn instantaneously. The pursuer wants to capture the evader by reducing his distance from player E to be less then ρ as early as possible.

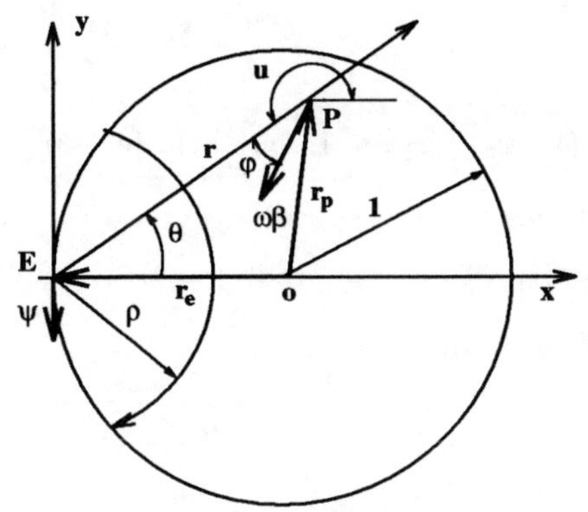

FIGURE 13.4. The reference frames for the Circular Wall Pursuit problem

We shall construct the solution for this problem by using two formulations simultaneously. One formulation will be in *fixed reference frames* yielding a 4-dimensional game model and the other will be in a *relative reference frame* which will turn to be a 2-dimensional game model.

The real frame

The origin of the reference frame is located at the center of the circular arena (at the point O in Figure 13.4). The position of player P at any instance is given by the two dimensional vector $\mathbf{r}_p$ and the position of player E at any instance is given by the two dimensional vector $\mathbf{r}_e$. The state vector of the system is a 4-dimensional vector which has two 2-dimensional vectors as its components.

$$\mathbf{z} = \begin{bmatrix} \mathbf{r}_p \\ \mathbf{r}_e \end{bmatrix}.$$

The relative frame

The origin of the relative reference frame is attached to player E and the $\theta = 0$ ray is confined to pass through the center of the arena (see Figure 13.4).

The state of the game is given by r, the distance between player P and player E, and θ, the angle of the direction of the location of player P relative to player E relative to the $\theta = 0$ ray (measured clockwise).

The game set:

In the *real frame*:

$$\mathcal{S}_f \equiv \{(\mathbf{r}_p, \mathbf{r}_e) \mid |\mathbf{r}_p| \leq 1, \; |\mathbf{r}_e| = 1, \; |\mathbf{r}_p - \mathbf{r}_e| \geq \rho\}.$$

In the *relative frame*:

$$\mathcal{S}_r \equiv \{(r, \theta) \mid r > \rho \; , \; r^2 + 1 - 2r\cos\theta \leq 1\}.$$

The target set:
In the *real frame*:

$$\mathcal{C}_f \equiv \{(\mathbf{r}_p, \mathbf{r}_e) \mid |\mathbf{r}_p| \leq 1,\ |\mathbf{r}_e| = 1,\ |\mathbf{r}_p - \mathbf{r}_e| = \rho\}.$$

In the *relative frame*:

$$\mathcal{C}_r \equiv \{(r, \theta) \mid r = \rho\ ,\ r^2 + 1 - 2r\cos\theta \leq 1\}.$$

The equations of motion:
In the *real frame*:

$$\overset{\circ}{\mathbf{r}}_p = w\boldsymbol{\beta}\ , \qquad |\boldsymbol{\beta}\,| = 1,$$

$$\overset{\circ}{\mathbf{r}}_e = \psi\hat{\mathbf{e}}_z \times \mathbf{r}_e\ , \quad |\psi| \leq 1.$$

($\hat{\mathbf{e}}_z$ is a unit vector perpendicular to the arena so that $\hat{\mathbf{e}}_z \times \mathbf{r}_e$ is in the plane of the arena).
Player P chooses the direction of the unit vector $\boldsymbol{\beta}$ and player E determines the speed and sense of his motion along the circumferene by choosing ψ.

In the *relative frame*:

$$\dot{r} = -w\cos\varphi + \psi\sin\theta,$$

$$\dot{\theta} = -\tfrac{1}{r}[w\sin\varphi + \psi(r - \cos\theta)]\ , \quad |\psi| \leq 1.$$

Player P chooses φ and player E determines the speed and sense of his motion along the circumferene by choosing ψ.

The outcome functional:
In the *real frame*:

$$G(\mathbf{r}_p, \mathbf{r}_e) = 0\ , \quad L(\mathbf{r}_p, \mathbf{r}_e, \boldsymbol{\beta}\ , \psi) = 1.$$

In the *relative frame*:

$$G(r, \theta) = 0\ , \quad L(r, \theta, \varphi, \psi) = 1.$$

We begin to solve the game problem in the *real frame*.

The partitioning of the target set
The normal to the target set is given by:

$$\mathbf{n}(\mathbf{r}_p, \mathbf{r}_e) = \begin{bmatrix} (\mathbf{r}_p - \mathbf{r}_e) \\ (\mathbf{r}_e - \mathbf{r}_p) \end{bmatrix},$$

and

$$\min_{\boldsymbol{\beta}}\ \max_{|\psi| \leq 1} \{w\boldsymbol{\beta} \cdot (\mathbf{r}_p - \mathbf{r}_e) + \psi\hat{\mathbf{e}}_z \times \mathbf{r}_e \cdot (\mathbf{r}_e - \mathbf{r}_p)\} < 0 \quad , \forall \begin{bmatrix} \mathbf{r}_p \\ \mathbf{r}_e \end{bmatrix} \in \mathcal{C}.$$

So:

$$UP \equiv \mathcal{C}_f .$$

The Candidate Optimal Control Laws

The equations of motion are separable and ME1 is:

$$\min_{\boldsymbol{\beta}} \max_{|\psi| \leq 1} \{w\boldsymbol{\beta} \cdot \mathbf{J_{r_p}} + \psi \hat{\mathbf{e}}_z \times \mathbf{r}_e \cdot \mathbf{J_{r_e}} + 1\}.$$

The candidate optimal control laws are:

$$\overset{*}{\psi} = \text{sign}\,(\hat{\mathbf{e}}_z \times \mathbf{r}_e \cdot \mathbf{J_{r_e}}) \,, \quad \text{for} \quad \hat{\mathbf{e}}_z \times \mathbf{r}_e \cdot \mathbf{J_{r_e}} \neq 0,$$

$$\overset{*}{\psi} = \text{ Any admissible } \psi \,, \quad \text{for} \quad \hat{\mathbf{e}}_z \times \mathbf{r}_e \cdot \mathbf{J_{r_e}} = 0,$$

$$\overset{*}{\boldsymbol{\beta}} \parallel -\mathbf{J_{r_p}}.$$

ME2 is:

$$-w|\mathbf{J_{r_p}}| + |\hat{\mathbf{e}}_z \times \mathbf{r}_e \cdot \mathbf{J_{r_e}}| + 1 = 0.$$

The Adjoint Equations

$$\overset{\circ}{\mathbf{J}}_{\mathbf{r_p}} (\mathbf{r}_p, \mathbf{r}_e) = 0,$$

$$\overset{\circ}{\mathbf{J}}_{\mathbf{r_e}} (\mathbf{r}_p, \mathbf{r}_e) = -\mathbf{J_{r_e}} \times \hat{\mathbf{e}}_z .$$

We have that along regular segments of the optimal trajectories $\mathbf{J_{r_p}}(\mathbf{r}_p, \mathbf{r}_e)$ is constant.

To find this constant adjacent to the UP we use the fact that the dimensionality of our problem is 2 and that in the 4-dimensional formulation of the problem there should exist 2 *invariant relations* involving the 4 components of the state vector.
Our system is symmetric and is not affected by rotations of the arena. The value function should therefore be a function only of the two variables (ϱ, ϑ):

$$J(\mathbf{r}_p, \mathbf{r}_e) = J(\varrho(\mathbf{r}_p, \mathbf{r}_e), \vartheta(\mathbf{r}_p, \mathbf{r}_e)),$$

where

$$\varrho = |\mathbf{r}_p - \mathbf{r}_e|,$$

$$\vartheta = \mathbf{r}_e \cdot \frac{(\mathbf{r}_p - \mathbf{r}_e)}{|\mathbf{r}_p - \mathbf{r}_e|}.$$

Obviously $\varrho = r$ and $\vartheta = \cos\theta$.
At the UP he have $\varrho = \rho$ and $J(\rho, \vartheta) = 0$ so:

$$\frac{\partial J}{\partial \vartheta}(\rho, \vartheta) = 0,$$

which implies that on the *UP*

$$\frac{\partial J}{\partial \mathbf{r}_p}(\mathbf{r}_p,\mathbf{r}_e) = \frac{\partial J}{\partial \varrho}(\rho,\vartheta)\frac{(\mathbf{r}_p-\mathbf{r}_e)}{\rho},$$

$$\frac{\partial J}{\partial \mathbf{r}_e}(\mathbf{r}_p,\mathbf{r}_e) = \frac{\partial J}{\partial \varrho}(\rho,\vartheta)\frac{(\mathbf{r}_e-\mathbf{r}_p)}{\rho}.$$

specializing ME2 to the *UP* we can now obtain that at the *UP*

$$\frac{\partial J}{\partial \mathbf{r}_p}(\mathbf{r}_p,\mathbf{r}_e) = \frac{(\mathbf{r}_p-\mathbf{r}_e)}{\rho(w-\sqrt{1-\left[\frac{\mathbf{r}_e\cdot(\mathbf{r}_p-\mathbf{r}_e)}{\rho}\right]^2})},$$

$$\frac{\partial J}{\partial \mathbf{r}_e}(\mathbf{r}_p,\mathbf{r}_e) = \frac{(\mathbf{r}_e-\mathbf{r}_p)}{\rho(w-\sqrt{1-\left[\frac{|\mathbf{r}_e\cdot(\mathbf{r}_p-\mathbf{r}_e)|}{\rho}\right]^2})}.$$

The Candidate Optimal Trajectories

The candidate optimal trajectory emanating backward from the *UP* in the *real frame* implies that the pursuer moves on a straight line towards E_f, the terminal position of player E.

The *τ-isochrone* in the *real frame* is a circle with a radius of magnitude $\rho + w\tau$ centered at E_f.

In the *relative frame* the *τ-isochrone* is also a circle with a radius of magnitude $\rho + w\tau$ but it is centered at E_τ where $\angle E_\tau O E_f = \tau$. (see Figure 13.5).

We parametrize the *UP* in the *relative frame* by the parameter θ:

$$x = \rho\cos\theta,$$

$$y = \rho\sin\theta.$$

The candidate optimal trajectory from $(x(\theta), y(\theta))$ is given by:

$$x = \cos\tau - (\rho + w\tau)\cos(\tau+\theta),$$

$$y = \sin\tau - (\rho + w\tau)\sin(\tau+\theta).$$

The dispersal surface

It is easy to see that for θ smaller then some $\tilde{\theta}$ the candidate optimal trajectories that emanate backward from the *UP* in the *relative frame* intersect the x-axis at a time $\tilde{\tau}(\theta)$:

$$\frac{1}{\sin(\theta+\tilde{\tau})} = \frac{\rho + w\tilde{\tau}}{\sin\tilde{\tau}}. \tag{13.74}$$

Examining the $\dot{\theta}$ equation in the *relative frame* for $\theta = 0$ we conclude that it is a case of *evader's dispersal surface* and it is player P who has to apply an *nonadmissible control law* of the type $\varphi = \varphi(\psi)$ on it.

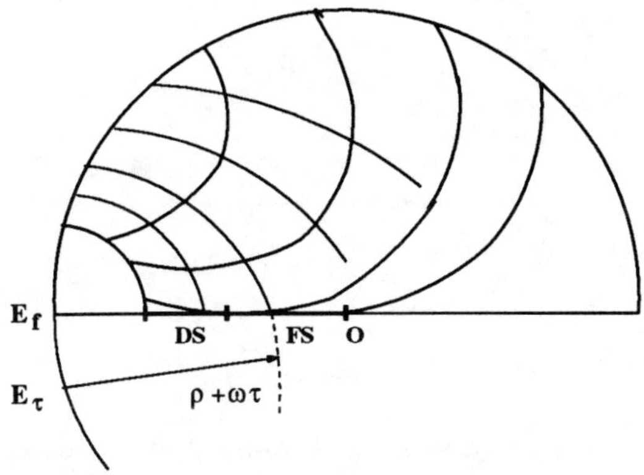

FIGURE 13.5. The isochrones in the relative frame

One particular trajectory arrives at the x-axis tangentially. this happens for $\theta = \tilde{\theta}$ where in addition to relation (13.74) the following is to be satisfied

$$-\cos\tilde{\tau} + (\rho + w\tilde{\tau})\cos(\tilde{\tau} + \tilde{\theta}) + \sin(\tilde{\tau} + \tilde{\theta}) = 0.$$

The *EDS* can not extend further backward from this point along the $\theta = 0$ axis.
We notice that the first field of candidate optimal trajectories fails to cover the game set.

The focal surface

Next we examine the possibility to obtain another field of optimal trajectories assuming that a *singular arc* emanates from the edge of the *EDS* and that tributaries arrive at it from both sides. We assume that it is player P who has to apply a *nonadmissible control law* of the type $\varphi = \varphi(\psi)$ along the singular arc.
At the edge of the *EDS* we have that $\dot{\theta} = 0$ on both arriving trajectories.
On the singular arc the problem:

$$\max H(r, \theta, \nabla J^{(i)}(r, \theta), \tilde{\varphi}^{(i)}(\psi), \psi),$$
$$\psi \in \{\psi \mid w\sin\varphi + \psi(r - \cos\theta) = 0\}.$$

is solved by both $\overset{*}{\psi}{}^{(i)}$ and $\tilde{\psi}^{(i)}$.
Here this problem is

$$\max_{\psi}\{-J_r^{(i)}(r, \theta)w[\sqrt{1 - [\tfrac{\psi(\cos\theta - r)}{w}]^2} + \tfrac{\psi}{w}\sin\theta]\}.$$

This problem has a unique solution (for ψ that provide that $\ddot{\theta}$ have the right sign for the paths that arrive from side-i).

So:

$$\overset{*}{\psi}^{(i)} = \tilde{\psi}^{(i)}.$$

The problem:

$$\min_{|\varphi| \leq 1} H(r, \theta, \nabla J^{(i)}(r, \theta), \varphi, \overset{*}{\psi}^{(i)}),$$

has a unique solution so:

$$\overset{*}{\varphi}^{(i)} = \tilde{\varphi}^{(i)}(\overset{*}{\psi}^{(i)}).$$

The tributaries to the singular arc join tangentially and we have a case of a *focal surface* along $\theta = 0$.

One can easily construct the tributaries in the *real frame* and transform them to the *relative frame.*

Notice that along the candidate *focal surface* $\theta = 0$:

$$\sin \varphi(\psi) = \psi \frac{(1-r)}{w}.$$

At the center O of the arena where $r = 1$ we have that $\varphi = 0$ at the tributaries from both sides regardless of ψ. Here player P does not need to use a law of the type $\varphi = \varphi(\psi)$ anymore and the singular arc cannot be extended further.

We still have a void in the game set uncovered by candidate optimal trajectories.

The Region of Multiple Choices

At $\mathbf{r_p} = 0$

$$\mathbf{J_{r_e}}(0, \mathbf{r_e}) = 0,$$

and we can deduce from the adjoint equations that along any regular optimal trajectory that emanate from the point O we have:

$$\overset{\circ}{\mathbf{J}}_{\mathbf{r_e}}(\mathbf{r_p}, \mathbf{r_e}) = 0,$$

and:

$$\mathbf{J_{r_e}}(\mathbf{r_p}, \mathbf{r_e}) = 0.$$

We recall that at the regular segments of candidate optimal trajectories in the *real frame* player P moves along straight lines. Hence the candidate optimal trajectories emanating backward from the center O of the arena imply a motion of player P along *radial lines* toward O while player E can choose any ψ.

$$\boldsymbol{\beta} \parallel \mathbf{r_p} \quad , \quad \overset{*}{\psi} = \text{any admissible } \psi.$$

We have thus succeeded to cover the game set with candidate optimal trajectories that satisfy the necessary conditions and the solution is complete.

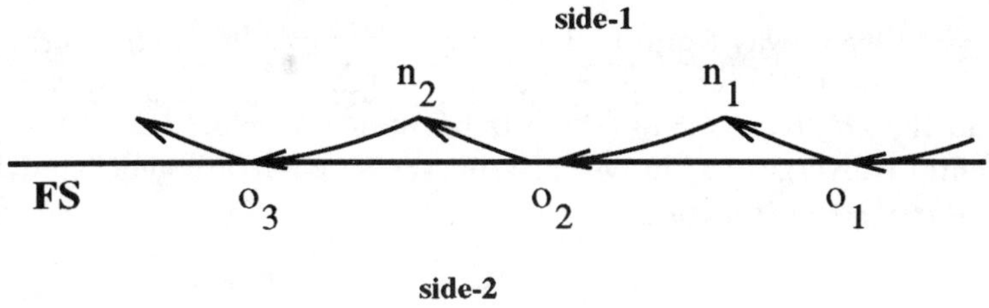

FIGURE 13.6. The chatter equivalent for the motion on the Focal Surface

13.6.1 The Chatter Equivalent

Let us examine what happens in our example if on the *focal surface* player P is not allowed to use a *nonadmissible control law* of the type $\varphi = \varphi(\psi)$. In the first quadrant, (denoted as *region 1*), we have:

$$0 < \theta < \tfrac{1}{2}\pi \quad , \overset{*(1)}{\psi} = 1 \quad , \sin \overset{*(1)}{\varphi} > 0,$$

and in the fourth quadrant, (denoted as *region 2*), we have:

$$-\tfrac{1}{2}\pi \leq \theta < 0 \quad , \overset{*(2)}{\psi} = -1 \quad , \sin \overset{*(2)}{\varphi} < 0.$$

At the point O_1 on the *FS* player P, who does not have an admissible strategy to prevent the crossing of the *FS*, might play $\overset{*(1)}{\varphi}$ in an *open-loop* fashion for a duration δ.
If player E plays $\overset{*(2)}{\psi}$ during that period we get:

$$\dot{\theta}(r, 0, \overset{*(1)}{\varphi}, \overset{*(2)}{\psi}) = -\frac{w \sin \overset{*(1)}{\varphi}}{r} + (\frac{1}{r} - 1) < 0,$$

and the trajectory leads to region-2 where the control $\overset{*(1)}{\varphi}$ may be nonoptimal for player P.
When the duration δ elapses and the trajectory reached the point n_1 inside region-2 (see Figure 13.6), player P resumes playing optimally in a *closed loop* fashion and by playing $\overset{*(2)}{\varphi}$ the trajectory leads back to the *FS* and joins it tangentially at the point o_2.
Here player P faces the same difficulties and may play open loop for another duration δ and so forth.

The result is a kind of a *chatter* where the segments generated by the pair $(\overset{*(2)}{\varphi}, \overset{*(2)}{\psi})$ lead towards the *FS* and the segments generated by the pair $(\overset{*(1)}{\varphi}, \overset{*(2)}{\psi})$ lead away from it. Both players apply *admissible control laws*. If we let $\delta \to 0$ the *chatter motion* converges to the singular arc on the *FS* and the *chatter velocity* (the mix between the velocities on the pair of segments) converges to the velocity on the singular arc generated by the *nonadmissible control law* $\varphi = \varphi(\psi)$

•

13.7 Problems

Problem 13.7.1 (The Circular Wall Pursuit problem)

Consider Example 13.6.1.

1. Show that in the *real frame* the motion of player P, corresponding to the *focal surface* is a motion on a circular arc with a radius of magnitude $\frac{1}{2}w$ that passes through the center of the arena and the edge of the *dispersal surface*.

•

Problem 13.7.2 (The Obstacle Tag problem)

Consider the *Obstacle Tag problem*. (See Problems 7.5.6 and 12.5.3).

1. The tributaries to the *safe contact* may reach the surface $\theta = \pi$ a part of which, due to symmetry, may constitute a candidate *dispersal surface*.
   Examine the case. Determine which player dominates.

2. Examine the feasibility that the curve $\frac{y^2}{x^2} = \frac{1}{w}$ on the surface $\theta = \pi$ (The locus where the tributaries to the *safe contact* are *tangent* to $\theta = \pi$) is the start of a candidate *focal surface*.
   Show that this can not be the case. (Hint: show that player P benefits if he uses $\tilde{b} = 0$, running towards the center instead of along a tangent to the circle).

•

Problem 13.7.3 (universal surfaces)

This problem deals with the construction of a candidate *universal surface*. Assume that:

$$\mathbf{f}(\mathbf{x}, \mathbf{u}, \mathbf{v}) = \mathbf{a}(\mathbf{x})\psi + \mathbf{b}(\mathbf{x}, \mathbf{u}, \mathbf{v_1}).$$

It follows that along regular optimal trajectories:

$$\dot{A}^{\psi} = \mathbf{J}_x[\mathbf{a}_x \mathbf{b} - \mathbf{b}_x \mathbf{a}]. \tag{13.75}$$

1. Derive a relation for computing $\tilde{\mathbf{f}}$ by taking the time derivative of equation 13.75 along a singular arc that lie on the *universal surface* (recalling that it should vanish there).

14

Corner Surfaces

14.1 Introduction

It is common in real life conflicts that we have to choose our decisions from a set of alternative decisions. Sometimes, as in the case of choosing which branch to follow at a road junction, the choice cannot be delayed and has to be done right on the spot. Other cases exist when one side can benefit from a *threat policy* where at each instance the choice can either be made or delayed.

Branch decisions are represented in differential games by *dispersal surfaces* (see chapter 11).
Threat policies are represented in differential games by *corner surfaces*.

We shall see that a segment of an optimal trajectory along which a threat of a delayed choice is maintained is a singular arc and that it lies on a singular surface across which $\mathbf{J}_x(\mathbf{x})$ is necessarily discontinuous.
Regular optimal trajectories leave the singular surface at each point representing the possibility to realize the threat.
We distinguish between two types of *corner surfaces*: *J. V. Breakwell's switch envelope* where the incoming regular optimal trajectories join tangentially and *R. Isaacs' equivocal surface* where the incoming regular optimal trajectories join transversely.
We give a characterization of the information structure on *corner surfaces* and show that in general the use of *nonadmissible strategies* by one player is necessary for a realization of an optimal motion along singular arc along a *corner surface*. We show that the use of *admissible* ϵ-optimal strategies lead to ϵ-optimal trajectories in the context of the *extended solution concept* and *chatter* controls.

14.2 Characterization of Corner Surfaces

We shall consider a singular surface as a *corner surface* if regular optimal trajectories *arrive* from one side (side-2) and *leave* to the other side (side-

1).

$$\mathbf{f}(\mathbf{x}, \overset{*}{\mathbf{u}}^{(1)}, \overset{*}{\mathbf{v}}^{(1)}) \cdot \mathbf{n}(\mathbf{x}) \leq 0,$$

$$\mathbf{f}(\mathbf{x}, \overset{*}{\mathbf{u}}^{(2)}, \overset{*}{\mathbf{v}}^{(2)}) \cdot \mathbf{n}(\mathbf{x}) \leq 0.$$

From the above definition of *corner surfaces* we exclude the specific case where the Hamiltonian $H(\mathbf{x}, \mathbf{J}_x, \mathbf{u}, \mathbf{v})$ is linear in components of the control vectors of *both* players and where *both* switch functions vanish at the junction of the tributaries to the *singular surface*.

Lemma 14.2.1

1. The function $\mathbf{J}_x(\mathbf{x})$ is discontinuous across a *corner surface*.
2. At a *corner surface*

$$\overset{*}{\mathbf{u}}^{(1)} \neq \overset{*}{\mathbf{u}}^{(2)}, \quad \overset{*}{\mathbf{v}}^{(1)} \neq \overset{*}{\mathbf{v}}^{(2)}. \tag{14.1}$$

Proof
Let us examine the following four cases:

1. Let

$$\begin{aligned} \mathbf{f}(\mathbf{x}, \overset{*}{\mathbf{u}}^{(1)}, \overset{*}{\mathbf{v}}^{(1)}) \cdot \mathbf{n}(\mathbf{x}) &= 0, \\ \mathbf{f}(\mathbf{x}, \overset{*}{\mathbf{u}}^{(2)}, \overset{*}{\mathbf{v}}^{(2)}) \cdot \mathbf{n}(\mathbf{x}) &< 0. \end{aligned} \tag{14.2}$$

We shall show that this cannot be a *corner surface*.
In this case the tributaries join the singular surface *transversely* implying that the Hamiltonian is linear in a component of a control vector and that the *switch function* for this component *vanishes* at the junction.
We assume, with no loss in generality, that the Hamiltonian is linear in a component of the control $\mathbf{v}$ of player E and that

$$\overset{*}{\mathbf{v}}^{(1)} \neq \overset{*}{\mathbf{v}}^{(2)}.$$

Obviously:

$$H[\mathbf{x}, J_x^{(2)}, \overset{*}{\mathbf{u}}^{(2)}, \overset{*}{\mathbf{v}}^{(2)}] = H[\mathbf{x}, J_x^{(2)}, \overset{*}{\mathbf{u}}^{(1)}, \overset{*}{\mathbf{v}}^{(1)}] = 0.$$

If

$$\overset{*}{\mathbf{u}}^{(1)} = \overset{*}{\mathbf{u}}^{(2)},$$

it is a case of a *regular switch surface* (see section 10.7).
and if

$$\overset{*}{\mathbf{u}}^{(1)} \neq \overset{*}{\mathbf{u}}^{(2)},$$

the problem

$$\min_{\mathbf{u} \in \mathcal{U}} H(\mathbf{x}, \mathbf{J}_x^{(2)}, \mathbf{u}, \mathbf{v}), \tag{14.3}$$

has two different solutions, $\overset{*}{\mathbf{u}}^{(1)}$ and $\overset{*}{\mathbf{u}}^{(2)}$ contrary to our assumption that the switch function for only one player vanishes on the tributaries at their junction to the *corner surface*.

2. Let

$$\begin{aligned} \mathbf{f}(\mathbf{x}, \overset{*}{\mathbf{u}}^{(1)}, \overset{*}{\mathbf{v}}^{(1)}) \cdot \mathbf{n}(\mathbf{x}) &= 0, \\ \mathbf{f}(\mathbf{x}, \overset{*}{\mathbf{u}}^{(2)}, \overset{*}{\mathbf{v}}^{(2)}) \cdot \mathbf{n}(\mathbf{x}) &= 0. \end{aligned} \tag{14.4}$$

The arguments for proving the assertions of the lemma for this case are the same as in the case of the *focal surface* (see section 13.4).

3. Let

$$\begin{aligned} \mathbf{f}(\mathbf{x}, \overset{*}{\mathbf{u}}^{(1)}, \overset{*}{\mathbf{v}}^{(1)}) \cdot \mathbf{n}(\mathbf{x}) &< 0, \\ \mathbf{f}(\mathbf{x}, \overset{*}{\mathbf{u}}^{(2)}, \overset{*}{\mathbf{v}}^{(2)}) \cdot \mathbf{n}(\mathbf{x}) &< 0. \end{aligned} \tag{14.5}$$

In this case, as in the first, tributaries arrive at the singular arc ***transversely*** implying that the Hamiltonian is linear in a component of the control vector and that the *switch function* for this component *vanishes* at the junction. We assume, with no loss in generality, that it is linear in a component of the control $\mathbf{v}$ of player E and that

$$\overset{*}{\mathbf{v}}^{(1)} \neq \overset{*}{\mathbf{v}}^{(2)},$$

If

$$\overset{*}{\mathbf{u}}^{(1)} = \overset{*}{\mathbf{u}}^{(2)},$$

it is a case of a *regular switch surface* (see section 10.7). and if

$$\overset{*}{\mathbf{u}}^{(1)} \neq \overset{*}{\mathbf{u}}^{(2)},$$

and

$$\nabla J^{(1)}(\mathbf{x}) = \nabla J^{(2)}(\mathbf{x}) = \nabla J(\mathbf{x}),$$

the problem

$$\min_{\mathbf{u} \in \mathcal{U}} H(\mathbf{x}, \mathbf{J}_x, \mathbf{u}, \mathbf{v}), \tag{14.6}$$

will have *different* solutions, $\overset{*}{\mathbf{u}}^{(1)}$ and $\overset{*}{\mathbf{u}}^{(2)}$ contrary to our assumption that the switch function for only one player vanishes at the junction of the tributaries to the *corner surface*.

4. Let

$$\mathbf{f}(\mathbf{x},\overset{*}{\mathbf{u}}^{(1)},\overset{*}{\mathbf{v}}^{(1)})\cdot\mathbf{n}(\mathbf{x}) < 0, \qquad \mathbf{f}(\mathbf{x},\overset{*}{\mathbf{u}}^{(2)},\overset{*}{\mathbf{v}}^{(2)})\cdot\mathbf{n}(\mathbf{x}) = 0. \tag{14.7}$$

If

$$\nabla J^{(1)}(\mathbf{x}) = \nabla J^{(2)}(\mathbf{x}),$$

and

$$\overset{*}{\mathbf{u}}^{(1)} \neq \overset{*}{\mathbf{u}}^{(2)}, \quad \overset{*}{\mathbf{v}}^{(1)} = \overset{*}{\mathbf{v}}^{(2)} = \overset{*}{\mathbf{v}}, \tag{14.8}$$

or

$$\overset{*}{\mathbf{v}}^{(1)} \neq \overset{*}{\mathbf{v}}^{(2)}, \quad \overset{*}{\mathbf{u}}^{(1)} = \overset{*}{\mathbf{u}}^{(2)} = \overset{*}{\mathbf{u}}, \tag{14.9}$$

neither player P nor player E need a nonadmissible control law, And if

$$\overset{*}{\mathbf{u}}^{(1)} \neq \overset{*}{\mathbf{u}}^{(2)}, \quad \overset{*}{\mathbf{v}}^{(1)} \neq \overset{*}{\mathbf{v}}^{(2)}, \tag{14.10}$$

the problems

$$\min_{\mathbf{u}\in\mathcal{U}} H(\mathbf{x},\mathbf{J}_x,\mathbf{u},\mathbf{v}), \tag{14.11}$$

and

$$\max_{\mathbf{v}\in\mathcal{V}} H(\mathbf{x},\mathbf{J}_x,\mathbf{u},\mathbf{v}), \tag{14.12}$$

will have nonunique solutions implying that the Hamiltonian is linear in components of the control vectors of both players and that both switch functions vanish.

If

$$\nabla J^{(1)}(\mathbf{x}) \neq \nabla J^{(2)}(\mathbf{x}),$$

and

$$\overset{*}{\mathbf{u}}^{(1)} \neq \overset{*}{\mathbf{u}}^{(2)}, \quad \overset{*}{\mathbf{v}}^{(1)} = \overset{*}{\mathbf{v}}^{(2)} = \overset{*}{\mathbf{v}}, \tag{14.13}$$

where the Hamiltonian is not linear in a component of $\mathbf{u}$ (If it where linear in $\mathbf{u}$ it would be a *switch surface*), the problems

$$\min_{\mathbf{u}\in\mathcal{U}} H(\mathbf{x},\mathbf{J}_x^{(2)},\mathbf{u},\mathbf{v}), \qquad \min_{\mathbf{u}\in\mathcal{U}} H(\mathbf{x},\mathbf{J}_x^{(1)},\mathbf{u},\mathbf{v}), \tag{14.14}$$

have each a *unique* solution and:

$$\begin{aligned} H(\mathbf{x},\mathbf{J}_x^{(1)},\overset{*}{\mathbf{u}}^{(1)},\overset{*}{\mathbf{v}}) = 0, \quad & H(\mathbf{x},\mathbf{J}_x^{(1)},\overset{*}{\mathbf{u}}^{(2)},\overset{*}{\mathbf{v}}) > 0, \\ H(\mathbf{x},\mathbf{J}_x^{(2)},\overset{*}{\mathbf{u}}^{(2)},\overset{*}{\mathbf{v}}) = 0, \quad & H(\mathbf{x},\mathbf{J}_x^{(2)},\overset{*}{\mathbf{u}}^{(1)},\overset{*}{\mathbf{v}}) > 0, \end{aligned} \tag{14.15}$$

it follows that

$$H(\mathbf{x}, \mathbf{J}_x^{(1)}, \overset{*}{\mathbf{u}}^{(1)}, \overset{*}{\mathbf{v}}) - H(\mathbf{x}, \mathbf{J}_x^{(2)}, \overset{*}{\mathbf{u}}^{(1)}, \overset{*}{\mathbf{v}}) = \beta(\mathbf{x})\mathbf{n}(\mathbf{x}) \cdot \mathbf{f}(\mathbf{x}, \overset{*}{\mathbf{u}}^{(1)}, \overset{*}{\mathbf{v}}) < 0,$$

and

$$H(\mathbf{x}, \mathbf{J}_x^{(1)}, \overset{*}{\mathbf{u}}^{(2)}, \overset{*}{\mathbf{v}}) - H(\mathbf{x}, \mathbf{J}_x^{(2)}, \overset{*}{\mathbf{u}}^{(2)}, \overset{*}{\mathbf{v}}) = \beta(\mathbf{x})\mathbf{n}(\mathbf{x}) \cdot \mathbf{f}(\mathbf{x}, \overset{*}{\mathbf{u}}^{(2)}, \overset{*}{\mathbf{v}}) > 0.$$

contradicting the definition which states that the trajectories at the two sides do not lead to opposite directions of the normal to the *corner surface*.

•

Lemma 14.2.2

On a *corner surface*

$$\begin{aligned} \beta(\mathbf{x})\mathbf{n}(\mathbf{x}) \cdot \mathbf{f}(\mathbf{x}, \overset{*}{\mathbf{u}}^{(1)}, \overset{*}{\mathbf{v}}^{(2)}) &< 0, \\ \beta(\mathbf{x})\mathbf{n}(\mathbf{x}) \cdot \mathbf{f}(\mathbf{x}, \overset{*}{\mathbf{u}}^{(2)}, \overset{*}{\mathbf{v}}^{(1)}) &> 0. \end{aligned} \tag{14.16}$$

Proof

The proof is left to the reader.

•

When $\beta(\mathbf{x}) > 0$ we have, by Lemma 14.2.2, that

$$\begin{aligned} \mathbf{f}(\mathbf{x}, \overset{*}{\mathbf{u}}^{(1)}, \overset{*}{\mathbf{v}}^{(2)}) \cdot \mathbf{n}(\mathbf{x}) &< 0, \\ \mathbf{f}(\mathbf{x}, \overset{*}{\mathbf{u}}^{(2)}, \overset{*}{\mathbf{v}}^{(1)}) \cdot \mathbf{n}(\mathbf{x}) &> 0. \end{aligned} \tag{14.17}$$

meaning that both the controls $\overset{*}{\mathbf{v}}^{(1)}$ and $\overset{*}{\mathbf{v}}^{(2)}$ of player E *fail* to satisfy the requirements of Lemma 10.5. Player E has to apply *nonadmissible control law* of the type $\mathbf{v} = \mathbf{v}(\mathbf{u})$ in order to realize the motion along the singular arc and satisfy the conditions of Lemma 10.5 for optimality. In this case we call the *corner surface evader's corner surface* and denote it by *ECS*. Similarly, if $\beta(\mathbf{x}) < 0$,

$$\begin{aligned} \mathbf{f}(\mathbf{x}, \overset{*}{\mathbf{u}}^{(1)}, \overset{*}{\mathbf{v}}^{(2)}) \cdot \mathbf{n}(\mathbf{x}) &> 0, \\ \mathbf{f}(\mathbf{x}, \overset{*}{\mathbf{u}}^{(2)}, \overset{*}{\mathbf{v}}^{(1)}) \cdot \mathbf{n}(\mathbf{x}) &< 0. \end{aligned} \tag{14.18}$$

and it is the case where player P will have to apply the *nonadmissible control law* of the type $\mathbf{u} = \mathbf{u}(\mathbf{v})$ in order to realize an optimal motion along the singular arc. In this case we call the *corner surface pursuer's corner surface* and denote it by *PCS*.

14.3 The Switch Envelope

If the optimal trajectories that arrive at the *corner surface* join *tangentially* the surface is a *switch envelope* (denoted as *SE*).
Consider the case of $\beta > 0$. this is the case where player E has to apply *nonadmissible control laws* of the type $\mathbf{v} = \mathbf{v}(\mathbf{u})$ in order to realize an optimal motion along the singular arc.
By Lemma 10.5.9 a tangential junction implies that

$$\overset{*}{\mathbf{u}}^{(2)} = \tilde{\mathbf{u}}, \qquad \overset{*}{\mathbf{v}}^{(2)} = \tilde{\mathbf{v}}(\tilde{\mathbf{u}}), \tag{14.19}$$

where $(\tilde{\mathbf{u}}, \tilde{\mathbf{v}}(\tilde{\mathbf{u}}))$ are the optimal controls along the singular arc on the *switch envelope surface*.
As in the cases of *universal surfaces* and *focal surfaces* the location of the *switch envelope* that may emanate from some candidate (n-2)-dimensional curve D is unknown ahead of time and we obtain it as part of the construction procedure. Often a *switch surface* is a continuation of a *dispersal surface*[1]. In general[2] we apply the adjoint equation for singular arcs, equation (10.55), to the *switch envelope* case (due to *P. Bernhard* [Ber77]) and integrate the following differential equations:[3]

$$\overset{\circ}{\mathbf{x}} = -\overset{*}{\mathbf{f}}^{(2)}(\mathbf{x}, \mathbf{J}_x), \tag{14.20}$$

$$\overset{\circ}{\mathbf{J}}_x^{(2)}(\mathbf{x}) = \frac{\partial \overset{*}{H}(\mathbf{x}, \mathbf{J}_x^{(2)})}{\partial x} + \Gamma(t)(\mathbf{J}_x^{(2)}(\mathbf{x}) - \mathbf{J}_x^{(1)}(\mathbf{x})), \tag{14.21}$$

subject to the requirement that the tributaries that arrive at the singular arc from field-2 join the *SE* tangentially:

$$(\mathbf{J}_x^{(1)}(\mathbf{x}) - \mathbf{J}_x^{(2)}(\mathbf{x}))\overset{*}{\mathbf{f}}^{(2)}(\mathbf{x}, \mathbf{J}_x). \tag{14.22}$$

To find $\Gamma(t)$ we take the time derivative of equation(14.22) along a singular arc that lies on the *SE*.
For example, for $L(\mathbf{x}, \mathbf{u}, \mathbf{v}) = 1$ and for *terminal cost games* we have that

[1]Notice that in such cases the seam between the *SE* and the *DS* is not necessarily the locus of points where regular optimal trajectories leave the *dispersal surface* tangentially (see *Breakwell* [Bre]) and in Problem 14.7.1 [Bre]).

[2]In 2-dimensional games we can often simplify the construction considerably by the use of ME2 for the adjacent region in order to find the magnitude of the adjoint vector at the junction with the tributaries.

[3]$\overset{*}{\mathbf{f}}^{(2)}$ is the velocity along the singular arc on the *SE*.

along the singular arc (using the notations of Lemma 10.6.3):

$$\frac{d}{dt}(\mathbf{J}_x^{(1)} \overset{*(2)}{\mathbf{f}}) = 0, \tag{14.23}$$

$$(\overset{*(2)}{\mathbf{f}})^T \mathbf{J}_{xx}^{(1)} \overset{*(2)}{\mathbf{f}} + \mathbf{J}_x^{(1)} \frac{d}{dt} \overset{*(2)}{\mathbf{f}} = 0, \tag{14.24}$$

We have that

$$\frac{d}{dt} \overset{*(2)}{\mathbf{f}} = \overset{*(2)}{\mathbf{f}_x} \overset{*(2)}{\mathbf{f}} + \overset{*(2)}{\mathbf{f}_{J_x}} \dot{\mathbf{J}}_x^{(2)}. \tag{14.25}$$

Substituting in equation (14.24) for $\frac{d}{dt} \overset{*(2)}{\mathbf{f}}$ from equation (14.25) and for $\dot{\mathbf{J}}_x^{(2)}$ from equation (14.21) recalling (Lemma 10.6.3) that

$$\overset{*(2)}{\mathbf{f}_{J_x}} = \overset{*\,(2)}{\mathbf{H}_{J_x J_x}}, \tag{14.26}$$

we obtain:

$$\begin{aligned} &\Gamma(t)(\mathbf{J}_x^{(1)})^T \overset{*\,(2)}{\mathbf{H}_{J_x J_x}} \mathbf{J}_x^{(1)} = \\ &(\overset{*(2)}{\mathbf{f}})^T \mathbf{J}_{xx}^{(1)} \overset{*(2)}{\mathbf{f}} + \mathbf{J}_x^{(1)} \overset{*(2)}{\mathbf{f}_x} \overset{*(2)}{\mathbf{f}} - \mathbf{J}_x^{(1)} \overset{*\,(2)}{\mathbf{H}_{J_x J_x}} (\overset{*\,(2)}{\mathbf{H}_x})^T. \end{aligned} \tag{14.27}$$

Example 14.3.1 (The Surveillance Evasion problem) *(See* J. Lewin *and* J. V. Breakwell *[LB75]).*

Let us consider the *surveillance evasion problem* again. In example 11.4.1 we denoted $\gamma = \sigma^{-1}$ and defined

$$f_1(\gamma) = \sqrt{1-\gamma^2} + \gamma(1 + \frac{1}{2}\pi - 2\sin^{-1}\gamma),$$

We have mentioned that for cases where β is in the range $1 < \beta < f_1(\gamma)$ and not to close to $f_1(\gamma)$, we encounter a *pursuer's dispersal surface* from the point $(\beta, 0)$ to a point D (see Figure 11.2) where a trajectory of field-2 (a straight ray from O) joins the *dispersal surface* tangentially.
Let us examine the possibility to construct another field of candidate optimal trajectories assuming that a *corner surface* emanates from the point D. It is easy to see that player E is not able to guarantee that a seam surface emanating from D is not crossed in a direction unfavorable to him. It is there for a case where player E utilizes an *nonadmissible control law* of the type $\psi(\phi, u)$ on the singular arc.
The problem:

$$\begin{aligned} &\min H(x, y, \nabla J^{(2)}, \varphi, u, \psi(\varphi, u)), \\ &|u| \le \sigma,\ |\varphi| \le 1 \end{aligned}$$

has a unique solution $u = 0$ so by Lemma 10.5.7

$$(\tilde{u}, \tilde{\varphi}) = (\overset{*}{u}^{(2)}, \overset{*}{\varphi}^{(2)}),$$

The problem:

$$\max_{\psi} H(x, y, \nabla J^{(2)}, \overset{*}{u}^{(2)}, \overset{*}{\varphi}^{(2)}, \psi)$$

has a unique solution so by Lemma 10.5.9

$$(\tilde{u}, \tilde{\varphi}) = (\overset{*}{u}^{(2)}, \overset{*}{\varphi}^{(2)}),$$

$$\tilde{\psi}(\tilde{u}, \tilde{\varphi}) = \overset{*}{\psi}^{(2)},$$

meaning that a construction of an *evader's switch envelope* from D might be feasible. Our example is 2-dimensional and the construction procedure can be simplified. In field-2:

$$\overset{*}{\varphi}^{(2)} = 0\,, \quad \overset{*}{\mathbf{u}}^{(2)} = 0,$$

$$\mathbf{f}(x, y, \overset{*}{\varphi}^{(2)}, \overset{*}{\mathbf{u}}^{(2)}, \overset{*}{\psi}^{(2)}) \parallel -\nabla J^{(2)}(x, y).$$

So by the *projection lemma*

$$\nabla J^{(1)}(x, y) \cdot \nabla J^{(2)}(x, y) = \nabla J^{(2)}(x, y) \cdot \nabla J^{(2)}(x, y), \tag{14.28}$$

and by ME2 in field-2:

$$|\nabla J^{(2)}(x, y)| = 1. \tag{14.29}$$

We can solve for $\nabla J^{(2)}(x, y)$ and obtain an expression for $\overset{*}{\psi}^{(2)}$ along the *switch envelope* in terms of the known components of $\nabla J^{(1)}(x, y)$ as well as an equation for the slope of the *switch envelope*. As we construct the *switch envelope* and its tributaries we must be aware that all the necessary conditions should be met. The continuation of the construction beyond the point G (see Figure 14.1) where the slope of the *switch envelope* and the slope of the outgoing trajectory of field-1 become identical does not provide a candidate optimal singular arc any more.
The solution can be carried further and completed by constructing a *focal surface* from G.

•

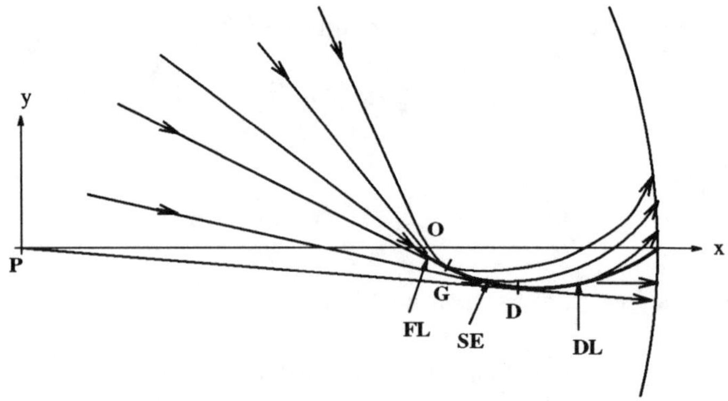

FIGURE 14.1. Switch envelope in the Surveillance Evasion game

14.4 Example of the Chatter Equivalent of a Switch Envelope Singular Arc

Let us examine what happens in our example if on the *switch envelope* player E is not allowed to utilize a *nonadmissible control law* of the type $\psi = \psi(\varphi)$.

Let us call the region above the *SE region-1* and the region below it *region-2*.

At the point O_1 on the *SE* player E, who does not have an admissible strategy to prevent the crossing of the *SE*, might play $\overset{*(1)}{\psi}$ in an *open-loop* fashion for a duration δ.

If player P plays $\overset{*(2)}{\varphi}$ during that period we get a trajectory that leads to region-2 where the control $\overset{*(1)}{\psi}$ may be *nonoptimal* for player E.

When the duration δ elapses and the trajectory reaches the point n_1 inside region-2 (see Figure 14.2), player E resumes playing optimally in a *closed loop* fashion and by playing $\overset{*(2)}{\psi}$ the trajectory leads back to the *SE* joins it tangentially at the point o_2.

Here player E faces the same difficulties and may play open loop for another duration δ and so forth.

The result is a *chatter* motion where the segments generated by the pair $(\overset{*(2)}{\varphi}, \overset{*(2)}{\psi})$ lead towards the *SE* and the segments generated by the pair $(\overset{*(1)}{\varphi}, \overset{*(2)}{\psi})$ lead away from it. Both players apply *information admissible control laws*.

If we let $\delta \to 0$ the *chatter motion* converges to the singular arc on the *SE* and the chatter-velocity (the mix between the velocities on the pair of segments) converges to the velocity along the singular arc that correspond

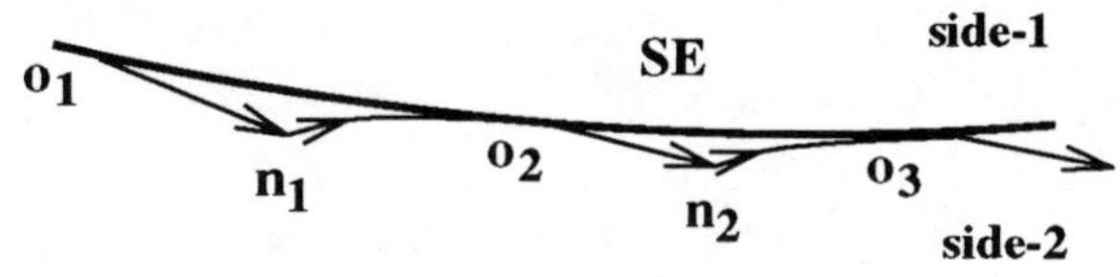

FIGURE 14.2. The chatter equivalent for the motion on the switch envelope

to the *nonadmissible control law* $\psi = \psi(\varphi)$

•

14.5 The Equivocal Surface

If the optimal trajectories that arrive at a *corner surface* join *transversely* the surface is an *equivocal survace* (denoted as ES).
Consider the case when $\beta > 0$. This is a case where player E has to apply *nonadmissible control laws* of the type $\mathbf{v} = \mathbf{v}(\mathbf{u})$ in order to realize an optimal motion along the singular arc.
By Lemma 10.5.10 it is necessary that the Hamiltonian $H(\mathbf{x}, \mathbf{J}_x, \mathbf{u}, \mathbf{v})$ will be linear in a component of the control $\mathbf{v}$ of player E and that the corresponding *switch function* will *vanish* on the optimal trajectories as they *arrive* at the *equivocal surface*. In the following we shall assume (for the sake of simplicity) that $H(\mathbf{x}, \mathbf{J}_x, \mathbf{u}, \mathbf{v})$ is linear only in one component of the velocity vector $\mathbf{v}$, the component ψ:

$$\begin{aligned} \mathbf{f}(\mathbf{x}, \mathbf{u}, \mathbf{v}) &= \mathbf{a}(\mathbf{x})\psi + \mathbf{b}(\mathbf{x}, \mathbf{u}, \mathbf{v_1}), \\ L(\mathbf{x}, \mathbf{u}, \mathbf{v}) &= c(\mathbf{x})\psi + d(\mathbf{x}, \mathbf{u}, \mathbf{v}_1). \end{aligned} \tag{14.30}$$

The switch function for $\overset{*}{\psi}$ is evidently:

$$A_\psi(\mathbf{x}, \mathbf{J}_x) = \mathbf{J}_x \cdot \mathbf{a}(\mathbf{x}) + c(\mathbf{x}). \tag{14.31}$$

As in the cases of *switch envelope* the location of the *equivocal surface* that may emanate from some candidate (n-2)-dimensional curve D is unknown ahead of time and we obtain it as part of the construction procedure.

In general[4] we have to apply the adjoint equations for singular arcs, (equation (10.55)), to the *equivocal surface* case (see *P. Bernhard* in [Ber77]) and integrate the following differential equations:

$$\overset{\circ}{\mathbf{x}} = \tilde{\mathbf{f}}, \tag{14.32}$$

and

$$\overset{\circ}{\mathbf{J}}{}_x^{(2)}(\mathbf{x}) = \frac{\partial \tilde{H}[\mathbf{x}, \mathbf{J}_x^{(2)}, \overset{*}{\mathbf{u}}{}^{(2)}, \tilde{v}(\overset{*}{\mathbf{u}}{}^{(2)})]}{\partial x} + \Gamma(t)(\mathbf{J}_x^{(2)}(\mathbf{x}) - \mathbf{J}_x^{(1)}(\mathbf{x})), \tag{14.33}$$

(where $\tilde{\mathbf{f}} = \mathbf{f}(\mathbf{x}, \overset{*}{\mathbf{u}}{}^{(2)}, \tilde{v}(\overset{*}{\mathbf{u}}{}^{(2)}))$ is the velocity along the singular arc on the ES), subject to the requirement that the *switch function* will vanish on the arriving optimal trajectories at their junction to the ES,

$$A_\psi(\mathbf{x}, \mathbf{J}_x^{(2)}) = 0, \tag{14.34}$$

and to the requirement, (due to the *projection lemma*, Lemma 10.2.1), that the singular arc lies on the ES:

$$(\mathbf{J}_x^{(1)} - \mathbf{J}_x^{(2)}) \cdot \tilde{\mathbf{f}} = 0. \tag{14.35}$$

Using ME2 we get that

$$\mathbf{J}_x^{(2)}(\mathbf{x}) \cdot \tilde{\mathbf{b}}(\mathbf{x}) + \tilde{d}(\mathbf{x}) = 0 \tag{14.36}$$

is equivalent to equation (14.34). To find a relation for computing $\Gamma(t)$, we take the time derivative of equation (14.36) along a singular arc that lie on the ES. Using the notations of Lemma 10.6.3 we get:

$$\mathbf{b}^T \mathbf{J}_{xx}^{(2)} \dot{J}_x^{(2)} + \mathbf{J}_x^{(2)} \mathbf{b}_x \tilde{\mathbf{f}} + \tilde{\mathbf{d}}_x \tilde{\mathbf{f}} = 0. \tag{14.37}$$

Substituting for $\dot{\mathbf{J}}_x^{(2)}$ from relation (14.33) we finally obtain:

$$\mathbf{J}_x^{(2)} \mathbf{b}_x \tilde{\mathbf{f}} + \tilde{\mathbf{d}}_x \tilde{\mathbf{f}} - \mathbf{b}^T \mathbf{J}_{xx}^{(2)} \tilde{H}_x^{(2)} + \Gamma(t) \mathbf{b}^T \mathbf{J}_{xx}^{(2)} (\mathbf{J}_x^{(1)} - \mathbf{J}_x^{(2)}) = 0. \tag{14.38}$$

14.6 Example of the Chatter Equivalent of the Equivocal Surface Singular Arc

Example 14.6.1 (The Conic Surveillance Evasion)

---

[4] In 2-dimensional games we can often simplify the construction considerably by the use of ME2 for the adjacent region in order to find the magnitude of the adjoint vector at the junction with the tributaries.

Let us consider the *Conic Surveillance Evasion* problem[5].
An *evader*, called player E, with a velocity of magnitude 1 (normalized) and an unlimited rate of turn try to escape from inside a *detection-set*, described as a 2-dimensional semi-infinite cone with half angle $\bar{\theta}$ in a planar reference frame (relative frame) attached to the velocity vector of a *pursuer*. The pursuer, player P, with a velocity of magnitude w and a minimum turn-radius of length 1 (normalized) tries to maintain the evader within the detection-set for maximum time.

The game-set:
In Cartesian form

$$\mathcal{S} \equiv \{(x, y) \mid y \geq 0 \; , \; |x| \leq y \tan \bar{\theta}\}.$$

In polar form

$$\mathcal{S} \equiv \{(r, \theta) \mid |\theta| \leq \bar{\theta}\}.$$

The target-set:
In Cartesian form

$$\mathcal{C} \equiv \{(x, y) \mid y \geq 0 \; , \; |x| = y \tan \bar{\theta}\}.$$

In polar form

$$\mathcal{C} \equiv \{(r, \theta) \mid |\theta| = \bar{\theta}\}.$$

The equations of motion:
In Cartesian form

$$\dot{x} = \sin \psi - yw\phi, \qquad |\phi| \leq 1,$$

$$\dot{y} = \cos \psi - w + xw\phi.$$

In polar form

$$\dot{r} = \cos(\psi - \theta) - w \cos \theta,$$

$$r\dot{\theta} = \sin(\psi - \theta) + w \sin \theta - rw\phi, \qquad |\phi| \leq 1.$$

Player P chooses ψ and player P chooses ϕ ($\phi = +1$ is a *full* turn to the right).

The outcome functional:

$$G(x, y) = 0 \quad , \; L(x, y, \phi, \psi) = 1.$$

---

[5]The problem was solved by *J. Lewin* and *G. J. Olsder* (see [LO79]).

Partitioning of the target-set
The normal to the branch $\theta = \bar{\theta}$ of the *target-set* is

$$\mathbf{n}(r, \bar{\theta}) = -\mathbf{e}_{\theta}.$$

We obtain that for

$$r < \sin\bar{\theta} + \frac{1}{w},$$

we have

$$\max_{\phi} \min_{\psi} -(\sin(\psi - \theta) + w\sin\theta - rw\phi) < 0,$$

so

$$UP \equiv \{(r, \theta) \mid r < \sin\bar{\theta} + \tfrac{1}{w},\ \theta = \bar{\theta}\},$$

$$NUP \equiv \{(r, \theta) \mid r > \sin\bar{\theta} + \tfrac{1}{w},\ \theta = \bar{\theta}\},$$

$$BUP \equiv \{(r, \theta) \mid r = \sin\bar{\theta} + \tfrac{1}{w},\ \theta = \bar{\theta}\}.$$

The candidate optimal control laws
The *Isaacs equation* (ME1) in Cartesian form is

$$\max_{\phi} \min_{\psi} \{J_x(\sin\psi - yw\phi) + J_y(\cos\psi - w + xw\phi) + 1\},$$

and in polar form

$$\max_{\phi} \min_{\psi} \{J_r[\cos(\psi\theta) - w\cos\theta] + \tfrac{1}{r}J_\theta[\sin(\psi - \theta) - rw\phi + w\sin\theta] + 1\},$$

and the optimal control laws are

$$\overset{*}{\phi} = \operatorname{sign}(xJ_y - yJ_x) = -\operatorname{sign} J_\theta,$$

$$(\sin\overset{*}{\psi}, \cos\overset{*}{\psi}) \parallel (-J_x, -J_y), \tag{14.39}$$

$$(\sin(\overset{*}{\psi} - \theta), \cos(\overset{*}{\psi} - \theta)) \parallel (-J_r, -\tfrac{1}{r}J_\theta).$$

The adjoint equations

$$\dot{J}_x = -J_y w \overset{*}{\phi},$$

$$\dot{J}_y = J_x w \overset{*}{\phi}, \tag{14.40}$$

or

$$\dot{J}_r = (J_\theta / r^2)[\sin(\psi - \theta) + w\sin\theta],$$

$$\dot{J}_\theta = -w(J_r \sin\theta + (J_\theta/r)\cos\theta) = -wJ_x. \tag{14.41}$$

For the terminal state $(r_f, \bar{\theta})$ that belong to the *usable part* (UP), we have

$$J_r(r_f, \bar{\theta}) = 0, \qquad J_\theta(r_f, \bar{\theta}) = -r_f/(1 + \sin\bar{\theta} - r_f w), \tag{14.42}$$

or

$$J_x(r_f \sin\bar{\theta}, r_f \cos\bar{\theta}) = \frac{-\cos\bar{\theta}}{1 + w\sin\bar{\theta} - r_f w}, \tag{14.43}$$

$$J_y(r_f \sin\bar{\theta}, r_f \cos\bar{\theta}) = \frac{\sin\bar{\theta}}{1 + w\sin\bar{\theta} - r_f w}. \tag{14.44}$$

The candidate optimal trajectories

Adjacent to the branch $\theta = \bar{\theta}$ of the *target-set* $J_\theta(r, \theta) < 0$ so on the UP,

$$\overset{*}{\phi} = +1,$$

$$\cos(\overset{*}{\psi} - \bar{\theta}) = 0,$$

$$\sin(\overset{*}{\psi} - \bar{\theta}) = 1.$$

We find that

$$\sin\overset{*}{\psi} = \cos(\bar{\theta} + w\overset{*}{\phi}\,\tau),$$

$$\cos\overset{*}{\psi} = -\sin(\bar{\theta} + w\overset{*}{\phi}\,\tau),$$

and for $\overset{*}{\phi} = +1$,

$$J_x(\tau) = -\frac{\cos(\bar{\theta} + w\tau)}{1 + w\sin\bar{\theta} - r_f w}, \tag{14.45}$$

$$J_y(\tau) = \frac{\sin(\bar{\theta} + w\tau)}{1 + w\sin\bar{\theta} - r_f w}, \tag{14.46}$$

$$x(\tau) = 1 - \cos(w\tau) + r_f \sin(\bar{\theta} + w\tau) - \tau\cos(\bar{\theta} + w\tau), \qquad y(\tau) = \sin(w\tau) + r_f \cos(\bar{\theta} + w\tau) + \tau\sin(\bar{\theta} + w\tau), \tag{14.47}$$

and

$$-J_\theta(\tau) = \frac{r\cos(\bar{\theta} - \theta + w\tau)}{1 + w\sin\bar{\theta} - r_f w}. \tag{14.48}$$

Relations (14.45)-(14.48) are not valid after $J_\theta(\tau)$ vanishes, i.e., for

$$\tau > \tau_s = (\frac{1}{2}\pi + \theta - \bar{\theta})/w. \tag{14.49}$$

The *isochrones* (surfaces of equal *time to go*) happen to be straight lines that are inclined at an angle $\bar{\theta} + w\tau$ to the y-axis and are tangent to a circle

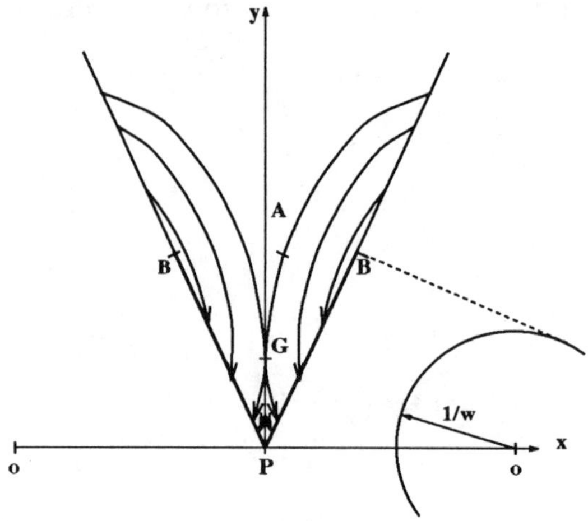

FIGURE 14.3. The first field of candidate optimal trajectories

of radius $\tau + \cos\bar{\theta}$, centered at O, which is the turn center of player P. The candidate optimal trajectories (14.47) have a clear geometric representation in the *real space*. Near termination, player P moves on a circular arc that corresponds to a full right turn $\overset{*}{\phi} = +1$, while player E moves along a straight line that is perpendicular to the target boundary at the instance of E's escape. Moreover, J_θ vanishes on such a path (indicating a possible switch in $\overset{*}{\phi}$), when the velocity of player E in the *real space* points directly at P.

As is often in the course of the solution of nontrivial differential games, the paths constructed backward from the *usable part* fail to fill the game-set with candidate optimal paths. This is an indication that a solution, if it exists, may exhibit discontinuities in $J(x, y)$ and its gradient across various manifolds in the game-set.

Figure 14.3 describes the candidate optimal trajectories (14.47) for a case with $\theta \leq \pi/4$. The point G is the unique point where a path that ends in the UP is tangent to the y-axis. Let us call this trajectory the *G-path*.

The equivocal surface

Let us first check if it is feasible that the *G-path* is a *switch surface* across which player P switches his choice of optimal control from $\overset{*}{\phi}^{(2)} = -1$ to $\overset{*}{\phi}^{(1)} = +1$.

Let A be a point on the *G-path*, further backwards than the point G, (see Figure 14.3). Figure 14.4 describes the situation at the point A by *vectograms*. Let us explain the figure in more detail.

The triangle ARL constitutes the *P-vectogram* where the segments $\overline{AR}$ and

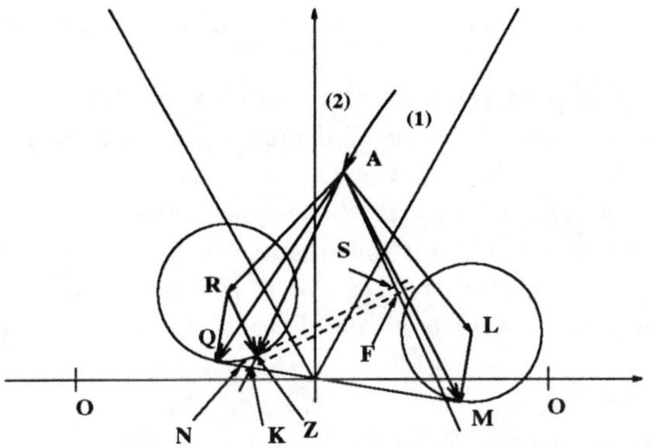

FIGURE 14.4. Vectograms for a point on the equivocal surface

$\overline{AL}$ correspond to $\overset{*}{\phi}= +1$ (right turn) and to $\overset{*}{\phi}= -1$ (left turn) respectively. The *E-vectogram* is a circle with a radius of magnitude 1.
In side-1 player P chooses $\overline{AR}$ and player E chooses $\overline{RZ}$ to be parallel to the direction of the normal to the *isochrone*. The segment $\overline{AZ}$ represents the resultant optimal velocity of the system at side-1 of the *G-path*. Notice that $\overline{AS}$ is the projection of this velocity on the normal to the isochrone and it represents the rate of decrease in *time to go* along the side-1 trajectory.
In side-2 player P chooses $\overline{AL}$ and player E may choose $\overline{LM}$. The resultant optimal velocity of the system at side-2 of the *G-path* is represented by $\overline{AM}$.

$$\overline{AZ} = \mathbf{f}(x, y, \overset{*}{\phi}{}^{(1)}, \overset{*}{\psi}{}^{(1)}),$$

$$\overline{AM} = \mathbf{f}(x, y, \overset{*}{\phi}{}^{(2)}, \overset{*}{\psi}{}^{(2)}),$$

$$\overline{AQ} = \mathbf{f}(x, y, \overset{*}{\phi}{}^{(2)}, \overset{*}{\psi}{}^{(1)}).$$

If the *G-path* is indeed a *switch surface* then upon arriving at A from side-2 both players change their choices of control.
Let us examine what happens if player E refrains from changing his choice. We recall that by the *Nash property* of the optimal solution it is necessary that player E will not benefit from such a unilateral deviation.
If player P chooses $\overline{AR}$ and player E chooses $\overline{RQ}$ the resultant velocity is represented by $\overline{AQ}$. This velocity leads back into side-2 where the choice $\overline{AR}$ is nonoptimal. Once he is back in side-2 player P again chooses $\overline{AL}$ and the resultant velocity leads back to side-2 of the *G-path* and so forth. We see that by refraining from changing the controls at the *G-path* player E can cause the path to *chatter* along the *G-path* with a resultant velocity

represented by $\overline{AK}$, (the *convex mix* of $\overline{AQ}$ and $\overline{AM}$ in the direction of the *G-path*).
Notice that $\overline{AF}$ is the projection of this velocity on the normal to the isochrone and it represents the rate of decrease in *time to go* along the side-1 trajectory. We conclude that the *G-path* cannot be a *switch surface* since $\overline{AF} > \overline{AS}$, (player E benefits from the deviation).
In order that the outcome will be indifferent to the option of player E to refrain from switching his control, player P will switch his control on a *singular switch surface* (the *equivocal surface*) which has the local direction of $\overline{AN}$. From Figure 14.4 we see that in this case the resultant *chatter velocity* along the *ES* (represented by $\overline{AN}$) and the resultant velocity in side-1 of the ,*ES* (represented by $\overline{AZ}$), have the *same* projection on the normal to the isochrone.

$$\overline{AN} = \mathbf{f}(x, y, \tilde{\phi}(\overset{*(2)}{\psi}), \overset{*(2)}{\psi}),$$

and

$$\mathbf{f}(x, y, \tilde{\phi}(\overset{*(2)}{\psi}), \overset{*(2)}{\psi}) \cdot \mathbf{n}(x, y) = \mathbf{f}(x, y, \overset{*(1)}{\phi}, \overset{*(1)}{\psi}) \cdot \mathbf{n}(x, y).$$

The formal construction of the equivocal surface

The problem

$$\max H(x, y, \nabla J^{(2)}, \tilde{\phi}(\psi), \psi),$$
$$\psi \in \{\psi | \mathbf{f}(x, y, \tilde{\phi}(\psi), \psi) \cdot \mathbf{n}(x, y) = 0\},$$

has a unique solution. We conclude that

$$\overset{*(2)}{\psi} = \tilde{\psi}.$$

We find $\tilde{\phi}(\tilde{\psi})$ by requiring that along the *ES*,

$$H(x, y, \nabla J^{(1)}, \tilde{\phi}(\tilde{\psi}), \tilde{\psi}) = 0,$$

$$1 - \tilde{\phi} = \frac{1 - - \sin(\bar{\theta} - \theta + w\tau)}{rw \cos(\bar{\theta} - \theta + w\tau)}.$$

We can obtain the right *ES*-curve by integrating the following differential equation starting from G:

$$\frac{dr}{d\theta} = \frac{-r(1 + w\cos\theta)\cos(\bar{\theta} - \theta + w\tau)}{(w\sin\theta - rw)\cos(\bar{\theta} - \theta + w\tau) + 1 - \sin(\bar{\theta} - \theta + w\tau)}.$$

We can find $\nabla J^{(2)}(x, y)$ and construct the tributaries to the *ES* by recalling that the switch function for $\overset{*}{\phi}$ vanishes at the junction,

$$xJ_y^{(2)}(x, y) - yJ_x^{(2)}(x, y) = 0,$$

together with ME2

$$H(x, y, \nabla J^{(2)}, \overset{*(2)}{\phi}, \overset{*(2)}{\psi}).$$

•

14.7 Problems

Problem 14.7.1 (The Obstacle Tag problem)

Consider the *Obstacle Tag* problem. (See Problems 7.5.6, 12.5.3 and 13.7.2).

1. Assume that the solution has a pair of *corner surfaces*, one in the North part $0 < \theta < \pi$ and its reflection in the South part, arriving tangent to each other at the 1-dimensional curve

$$\theta = \pi,$$

$$w[1 - \sqrt{1 - \tfrac{1}{x^2}}] = [1 - \sqrt{1 - \tfrac{1}{y^2}}] = s, \quad 0 < s < 1.$$

Denote the North primary field by the superscript (1) and the field of tributaries to the North *corner surface* by (2).
Examine the nature of the relevant Hamiltonians and show that a candidate *corner surface* should here be a *switch envelope*.

2. Write expressions for $\overset{*}{b}^{(2)}, \overset{*}{c}^{(2)}$.

3. Write the singular adjoint equations for the singular arc and derive the relation for $\Gamma(t)$:

$$\begin{aligned} \Gamma(t)\mathbf{J}_x^{(1)T} \overset{*}{\mathbf{H}}{}^{(2)}_{J_x J_x} \mathbf{J}_x^{(1)} = \; & \dot{\mathbf{x}}^{(2)T}\mathbf{J}_{xx}^{(1)}\dot{\mathbf{x}}^{(2)} + \\ & + \mathbf{J}_x^{(1)T}\dot{\mathbf{x}}_x^{(2)}\dot{\mathbf{x}}^{(2)} - \; \mathbf{J}_x^{(1)T} \overset{*}{\mathbf{H}}{}^{(2)}_{J_x J_x} \overset{*}{\mathbf{H}}{}^{(2)T}_x . \end{aligned} \tag{14.50}$$

4. Denoting (as was done by *J. V. Breakwell* [Bre]) the directions of player P and player E arriving at and departing from a *switch envelope* by $\overset{*}{\phi}{}^{(2)}, \overset{*}{\psi}{}^{(2)}$ and $\overset{*}{\phi}{}^{(1)}, \overset{*}{\psi}{}^{(1)}$ show that the arrival Hamiltonian is

$$H[\mathbf{x}, \nabla J^{(2)}, \phi, \psi] = 1 + A\cos(\psi - \overset{*}{\psi}{}^{(2)}) - (1 + A)\cos(\phi - \overset{*}{\phi}{}^{(2)}), \tag{14.51}$$

where

$$A = \sqrt{(J_x^{(2)})^2 + \left(\frac{J_\theta^{(2)}}{y}\right)^2},$$

and that the departure Hamiltonian is:

$$H[\mathbf{x}, \nabla J^{(1)}, \phi, \psi] = 1 + \tfrac{1}{w-1}\cos(\psi - \overset{*}{\psi}{}^{(1)}) - \tfrac{w}{w-1}\cos(\phi - \overset{*}{\phi}{}^{(1)}). \tag{14.52}$$

5. What else should be checked to validate the construction of the *switch envelope*?

•

Problem 14.7.2 (The Servo problem) *(see P. Bernhard in [Ber77])*

Consider Example 4.2.1.

1. Construct the primary fields of optimal trajectories emanating backwards from the two parts of the *usable-part*.

2. Show that in order to prevent the crossing of the $D^{(2)}$ *barrier* from its (1)-side to its (2)-side near its $UP^{(2)}$ edge, player E must use a control law of the type $v = v(x, u)$.

3. Show that player P can benefit at points on the (1)-side of the *barrier* $D^{(2)}$ near the $UP^{(2)}$ by deviating from $\overset{*}{u} = -1$ to $u = 1$.

4. Try a construction of a *corner surface* to resolve this issue. Show that it is a candidate *equivocal surface* and that it starts at the point $C^{(2)}$ on the *barrier* $D^{(2)}$ where:

$$x_c = \sqrt{\frac{2-2q}{q}},$$

$$y_c = q - 1 + \sqrt{2q(1-q)}.$$

5. Write an expression for the slope of the *equivocal surface* as a function of the parameters of the (1)-field, (The field of optimal trajectories emanating backwards from the *UP* of the $y = -1$ part of the *target-set*).

6. Show that the *ES* joins the $D^{(2)}$ *barrier* tangentially.

•

15

The Envelope Barrier

15.1 Introduction

In chapter 8 we saw that a procedure for the construction of a *semipermeable surface* through a given (n-2)-dimensional curve R in a n-dimensional game set is feasible if the *barrier paths* join R *nontangentially*[1] .
In this chapter we shall describe how in some games we can identify specific curves R belonging to particular surfaces S through which a special kind of SPS (which *R. Isaacs* called *envelope barrier*), with *barrier paths* that are *tangent* to the curve R can be constructed.

15.2 The Envelope Barrier

15.2.1 Dominated Surfaces

Suppose that at a point $\mathbf{x}$ player E has for a certain direction $\mathbf{n}(\mathbf{x})$ a control law $\mathbf{v} = \mathbf{v}(\mathbf{u})$ such that for any $\mathbf{u} \in \mathcal{U}$:

$$\mathbf{n}(\mathbf{x}) \cdot \mathbf{f}(\mathbf{x}, \mathbf{u}, \mathbf{v}(\mathbf{u})) = \mathbf{0}. \tag{15.1}$$

This evidently represents a case where at $\mathbf{x}$ the direction $\mathbf{n}(\mathbf{x})$ is *dominated kinematically* by player E.
A surface S is said to be *kinematically dominated* (denoted as KDS) by player E if relation (15.1) holds for all points $\mathbf{x} \in S$ [2]

15.2.2 Characterization of Envelope Barriers

Suppose that the surface S is a KDS dominated by player E. The velocity vector $\mathbf{f}(\mathbf{x}, \mathbf{u}, \mathbf{v}(\mathbf{u}))$ generates trajectories that lie on S and can be considered to belong to a *reduced order* (n-1)-dimensional game-system defined on S.

---

[1] Recall from section 8.2 that when the velocity vector $\mathbf{f}$ is tangent to R relation (8.14) is linearly dependent of relations (8.13) and we cannot obtain the initial values needed for the backward integration of $\frac{\partial \mathbf{n}[\mathbf{y}(\mathbf{l},\tau)]}{\partial \tau}$.

[2] The reader should be aware of the difference in information structure between a KDS and a SPS. On a KDS one player, say player E, can prevent the other player from crossing the KDS in *both* senses of the normal $\mathbf{n}(\mathbf{x})$. Notice that in order to do so he has to employ a *nonadmissible control law* $\mathbf{v} = \mathbf{v}(\mathbf{u})$.

Let $\tilde{\mathbf{v}}(\mathbf{u})$ be the solution of

$$\begin{array}{l} \max H(\mathbf{x},\mathbf{n},\mathbf{u},\mathbf{v}) \\ \mathbf{v} \in \{\mathbf{v} \mid \mathbf{v} \in \mathcal{V},\ \mathbf{n}(\mathbf{x}) \cdot \mathbf{f}(\mathbf{x},\mathbf{u},\mathbf{v}) = \mathbf{0}\}. \end{array} \tag{15.2}$$

If $\tilde{\mathbf{u}}$ solves the problem

$$\begin{array}{l} \min H(\mathbf{x},\mathbf{n},\mathbf{u},\tilde{\mathbf{v}}(\mathbf{u})) = 0. \\ \mathbf{u} \in \mathcal{U} \end{array} \tag{15.3}$$

The velocity vector $\mathbf{f}(\mathbf{x},\tilde{\mathbf{u}},\tilde{\mathbf{v}}(\tilde{\mathbf{u}}))$ generates curves D lying on the surface S that can be considered as semipermeable surfaces with respect to the *reduced game system*.

Let us examine the possibility of constructing a *semipermeable surface* R tangent to the surface S through the curve D.

Let us parametrize the curve D as we did in section 8.2 by

$$\mathbf{x} = \mathbf{h}(\mathbf{l}),$$

where $\mathbf{l}$ has $(n-2)$ components.

We can obtain $(n-2)$ relations between the values of the components of the normal $\mathbf{n}$ to the surface R at points $\mathbf{x}$ that belong to D through the requirement that R belongs to D:

$$\mathbf{n} \cdot \mathbf{h}_l = 0. \tag{15.4}$$

Notice that we cannot use ME2 for deriving another relation because the the velocity vector $\mathbf{f}(\mathbf{x},\tilde{\mathbf{u}},\tilde{\mathbf{v}}(\tilde{\mathbf{u}}))$ is tangent to the curve D.

The pair $(\overset{*}{\mathbf{u}},\overset{*}{\mathbf{v}})$ of *barrier controls* on R should satisfy:

$$\begin{array}{ll} \min & \max H(\mathbf{x},\mathbf{n},\mathbf{u},\mathbf{v}) = 0 = H(\mathbf{x},\mathbf{n},\overset{*}{\mathbf{u}},\overset{*}{\mathbf{v}}). \\ \mathbf{u} & \mathbf{v} \end{array} \tag{15.5}$$

So evidently by relation (15.5):

$$H(\mathbf{x},\mathbf{n},\overset{*}{\mathbf{u}},\tilde{\mathbf{v}}(\overset{*}{\mathbf{u}})) \le H(\mathbf{x},\mathbf{n},\overset{*}{\mathbf{u}},\overset{*}{\mathbf{v}}) = 0, \tag{15.6}$$

and by relation (15.10):

$$H(\mathbf{x},\mathbf{n},\overset{*}{\mathbf{u}},\tilde{\mathbf{v}}(\overset{*}{\mathbf{u}})) \ge H(\mathbf{x},\mathbf{n},\tilde{\mathbf{u}},\tilde{\mathbf{v}}(\tilde{\mathbf{u}})) = 0. \tag{15.7}$$

We infer that both $\overset{*}{\mathbf{v}}$ and $\tilde{\mathbf{v}}(\overset{*}{\mathbf{u}})$ solve the problem

$$\begin{array}{l} \max H(\mathbf{x},\mathbf{n},\overset{*}{\mathbf{u}},\mathbf{v}). \\ \mathbf{v} \in \mathcal{V} \end{array} \tag{15.8}$$

Lemma 15.2.1

If the problem

$$\max_{\mathbf{v} \in \mathcal{V}} H(\mathbf{x}, \mathbf{n}, \overset{*}{\mathbf{u}}, \mathbf{v}), \tag{15.9}$$

has a unique solution then the *barrier paths* of R are tangent to the surface S at D.

Proof

The proof follows immediately from the fact that

$$\mathbf{n}(\mathbf{x}) \cdot \mathbf{f}(\mathbf{x}, \overset{*}{\mathbf{u}}, \overset{*}{\mathbf{v}}) = \mathbf{n}(\mathbf{x}) \cdot \mathbf{f}(\mathbf{x}, \overset{*}{\mathbf{u}}, \tilde{v}(\overset{*}{\mathbf{u}})) = 0.$$

•

Lemma 15.2.2

If both the problem

$$\min_{\mathbf{u} \in \mathcal{U}} H(\mathbf{x}, \mathbf{n}, \mathbf{u}, \tilde{\mathbf{v}}(\mathbf{u})) = 0, \tag{15.10}$$

and the problem

$$\max_{\mathbf{v} \in \mathcal{V}} H(\mathbf{x}, \mathbf{n}, \overset{*}{\mathbf{u}}, \mathbf{v}), \tag{15.11}$$

have unique solutions then the *barrier paths* of R are tangent to the surface S and to the curve D as well.

Proof

The proof follows from the fact that we have:

$$\mathbf{f}(\mathbf{x}, \overset{*}{\mathbf{u}}, \overset{*}{\mathbf{v}}) = \mathbf{f}(\mathbf{x}, \tilde{\mathbf{u}}, \tilde{\mathbf{v}}(\tilde{\mathbf{u}})). \tag{15.12}$$

•

The information that at points $\mathbf{x}$ that belong to the curve D:

$$\mathbf{n}(\mathbf{x}) \cdot \mathbf{f}(\mathbf{x}, \overset{*}{\mathbf{u}}, \overset{*}{\mathbf{v}}) = 0,$$

enable us, by ME1, to obtain an independent relation involving the missing components of the normal $\mathbf{n}(\mathbf{x})$ at points that belong to the curve D and to realize a backward construction of a *semipermeable surface* R from the curve D.

We conclude that if in a game model we have a KDS, and we can construct a curve D that is semipermeable with respect to the corresponding *reduced order system* restricting the motion to stay on the KDS and if Lemma 15.2.2 holds we can construct an *envelpoe barrier* through the curve D.

15.3 Examples

Example 15.3.1 (The Lion and Man problem) *[Lew86,Fly73,Fly74]*

Let us consider the *Lion and Man* problem[3] . A *man*, armed with a gun that has an effective range of magnitude ρ, chases a *lion*. The chase takes place in a circular arena with radius 1 where $(1 > \rho)$. The speed of *lion* is v_L is greater than the speed of v_M of *man*. Both are capable of instantaneous turns. The *lion* is clever and in spite of the fact that he is faster, $(v_L > v_E)$, he tries to evade *man*.
We want to know under what circumstances can *lion* guarantee escape.
Let us approach this game by using two formulations simultaneously. One formulation will be in a fixed reference frame, which we shall call the *real frame*, yielding a 4-dimensional game model and the other will be in a *relative reference frame* which will turn to be a 3-dimensional game model.

The real frame

The origin of the *fixed reference frame* is located at the center of the circular arena (at the point O in Figure 15.1). The position of player P in the arena at any instance is given by $\mathbf{r}_p$ and the position of player E at any instance is given by $\mathbf{r}_e$.
The state vector of the system is the 4-dimensional vector

$$\mathbf{z} = \begin{pmatrix} \mathbf{r}_p \\ \mathbf{r}_e \end{pmatrix}.$$

The relative frame

The origin of the *relative reference frame* is attached to player E and the $\theta = 0$ ray is confined to pass through the center of the arena (see Figure 15.1).
The state of the game is given by r, the distance between player P and player E, θ, the angle of the direction of the location of player P relative to player E with respect to the $\theta = 0$ ray (measured counterclockwise) and r_e, the distance of player E from the center of the arena.

The game-set:
In the *real frame*:

$$\mathcal{S} \equiv \{(\mathbf{r}_p, \mathbf{r}_e) \mid |\mathbf{r}_p| \leq 1,\ |\mathbf{r}_e| \leq 1,\ |\mathbf{r}_p - \mathbf{r}_e| \geq \rho\}.$$

[3]The problem was first discussed by *J. Flynn* (see [Fly73] and [Fly74]) who gave bounds for the critical capture radius using direct methods. *J. Lewin* treated the problem with the methods of the theory of differential games and obtained the rigorous value for the critical radius (see [Lew86]). *J. V. Breakwell* analyzed the time-optimal pursuit problem and presented a computed solution for a specific case in [Bre89].

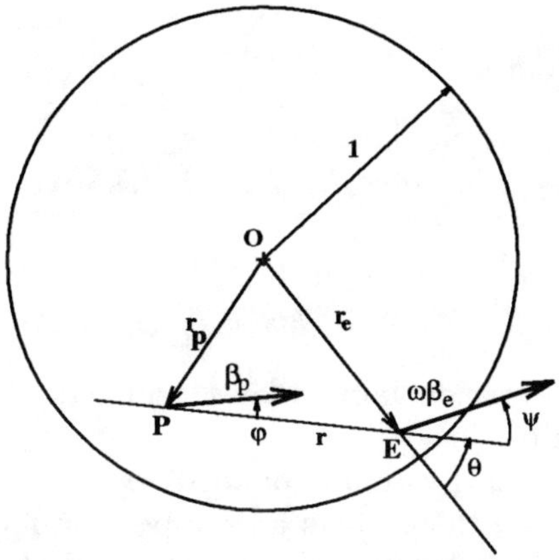

FIGURE 15.1. Reference frames for the Lion and Man problem

In the *relative frame*:

$$\mathcal{S} \equiv \{(r, \theta, r_e) \mid 0 \leq r_e \leq 1, \quad r \geq \rho, \quad r_e^2 + r^2 - 2r_e r \cos\theta \leq 1\}.$$

The target-set:
In the *real frame*:

$$\mathcal{S} \equiv \{(\mathbf{r}_p, \mathbf{r}_e) \mid |\mathbf{r}_p| \leq 1, \ |\mathbf{r}_e| \leq 1, \ |\mathbf{r}_p - \mathbf{r}_e| = \rho\}.$$

In the *relative frame*:

$$\mathcal{C} \equiv \{(r, \theta, r_e) \mid 0 \leq r_e \leq 1, \quad r = \rho, \quad r_e^2 + r^2 - 2r_e r \cos\theta \leq 1\}.$$

The equations of motion:
In the *real frame*:

$$\overset{\circ}{\mathbf{r}}_p = \boldsymbol{\beta}_p \quad , \qquad |\boldsymbol{\beta}_p| = 1,$$

$$\overset{\circ}{\mathbf{r}}_e = w\boldsymbol{\beta}_e \quad , \qquad |\boldsymbol{\beta}_e| = 1.$$

$w > 1$ is a parameter.
Player P chooses the direction of the unit vector $\boldsymbol{\beta}_p$ and player E chooses the direction of the unit vector $\boldsymbol{\beta}_e$.
In the *relative frame*:

$$\dot{r} = w\cos\psi - \cos\varphi,$$

$$r\dot{\theta} = w\sin\psi - \sin\varphi - \tfrac{wr}{r_e}\sin(\psi + \varphi),$$

$$\dot{r_e} = w\cos(\psi + \varphi).$$

Player P chooses φ and player E chooses ψ

The outcome functional:
In the *real frame*:

$$G(\mathbf{r}_p, \mathbf{r}_e) = 0 \ , \quad L(\mathbf{r}_p, \mathbf{r}_e, \boldsymbol{\beta}_p \ , \boldsymbol{\beta}_e \ ) = 1.$$

In the *relative frame*:

$$G(r, \theta, r_e) = 0, \qquad L(r, \theta, r_e, \varphi, \psi) = 1.$$

This is a pure pursuit evasion game. Player E prefers nontermination over any quantitative result.
We want to investigate the possibility of an *escape zone*. If, for some values of the parameters (ρ, w), there exists an *escape zone* for player E, then it should necessarily be bounded by a closed *semipermeable surface*. In many cases, the *barriers* emanate from the *BUP* [4]. So, we approach the problem by trying to construct an *envelope barrier* emanating from the *BUP*.

The partitioning of the target-set
Let us start the analysis in the *relative frame*.
The normal to the target-set pointing into the game-set is given by

$$\mathbf{n}(\rho, \theta, r_e) = -\hat{\mathbf{e}}_r,$$

where $\hat{\mathbf{e}}_r$ is a unit vector in the r_e direction.
We find that:

$$\min_{\varphi} \ \max_{\psi} \{-(w \cos \psi - \cos \phi) r\} < 0 \ , \qquad \text{only for } (\rho, \theta, r_e) \text{ with } r_e = 1. \text{ and } w \cos(\psi + \theta) < 0$$

So

$$UP \equiv \{(r, \theta, r_e) \in C, \ r_e = 1, \ |\theta| < \sin^{-1} w^{-1} \ \},$$

$$NUP \equiv \{(r, \theta, r_e) \in C, \ r_e < 1, \ \text{and } r_e = 1, \ |\theta| > \sin^{-1} w^{-1} \ \},$$

$$BUP \equiv \{(r, \theta, r_e) \in C, \ r_e = 1, \ |\theta| = \sin^{-1} w^{-1} \ \}.$$

The candidate control laws
Let us now proceed in the *real frame*.
ME1 is

$$\min_{\boldsymbol{\beta}_p} \ \max_{\boldsymbol{\beta}_e} \mathbf{n}_{\mathbf{r_p}} \cdot \boldsymbol{\beta}_p + w\mathbf{n}_{\mathbf{r_e}} \cdot \boldsymbol{\beta}_e \ ,$$

and the candidate optimal control laws are:

$$\overset{*}{\boldsymbol{\beta}}_p \| -\mathbf{n}_{\mathbf{r_p}}, \quad \overset{*}{\boldsymbol{\beta}}_e \| \ \mathbf{n}_{\mathbf{r_e}}.$$

---

[4]The *BUP* does not necessarily have to belong to the *barrier*, see Example 12.4.1 for a simple case when it does not!

The adjoint equations

$$\overset{\circ}{\mathbf{n}}_{\mathbf{r_p}} = 0,$$

$$\overset{\circ}{\mathbf{n}}_{\mathbf{r_e}} = 0.$$

We can easily conclude that the motions in the real frame that correspond to the *regular barrier paths* are motions along straight lines by both players.

The envelope barrier

It is easy to see that the surfaces $r = a$ are *kinematically dominated* by player E. Player E can choose $\psi(\varphi)$ so that $w \cos \psi(\varphi) = \cos \varphi$ and thus maintain $\dot{r} = 0$ at each point of the surface $r = a$.

Let us now construct a semipermeable curve D with respect to a reduced order system (θ, r_e) restricted to the surface $r = a$. The *Hamiltonian* for the reduced system is:

$$\bar{H}(\theta, r_e, \bar{\mathbf{n}}, \varphi) = \bar{n}_\theta(\theta, r_e)\dot{\theta}(\theta, r_e, \varphi, \psi(\varphi)) + \bar{n}_{r_e}(\theta, r_e)\dot{r_e}(\theta, r_e, \varphi, \psi(\varphi)).$$

The *barrier control* $\tilde{\varphi}$ for the reduced order system solves:

$$\min_{\varphi} \bar{H}(\theta, r_e, \bar{\mathbf{n}}, \varphi),$$

and satisfies:

$$\frac{d\bar{H}(\theta, r_e, \bar{\mathbf{n}}, \varphi)}{d\varphi} = 0. \tag{15.13}$$

ME2 gives:

$$\bar{n}_\theta(\theta, r_e)\dot{\theta}(\theta, r_e, \tilde{\varphi}, \psi(\tilde{\varphi})) + \bar{n}_{r_e}(\theta, r_e)\dot{r_e}(\theta, r_e, \tilde{\varphi}, \psi(\tilde{\varphi})) = 0. \tag{15.14}$$

For the system of equation (15.13) and equation (15.14) to have a nontrivial solution we must have

$$\frac{d\dot{\theta}}{d\varphi}\dot{r}_e - \dot{\theta}\frac{d\dot{r}_e}{d\varphi} = 0. \tag{15.15}$$

Denoting μ as

$$\mu = \theta + \cos^{-1}\left(\frac{\cos\tilde{\varphi}}{w}\right),$$

we can write equation (15.15) for $\tilde{\varphi}$ as

$$\begin{aligned} &w \cos\mu[(w - \tfrac{1}{w})\cos\tilde{\varphi} - \tfrac{wr}{r_e}\cos\mu] + \\ &+ w\sin\mu[w\sqrt{1 - w^2\cos^2\tilde{\varphi}} - \sin\tilde{\varphi} - \tfrac{wr}{r_e}\sin\mu] = 0. \end{aligned} \tag{15.16}$$

We can solve equation (15.16) for $\tilde{\varphi}$ as a function of θ and get $\tilde{\psi}(\tilde{\varphi})$ as well. We have now an equation for the slope $\frac{dr_e}{d\theta}$ from which the equation $r_e = r_e(\theta)$ for the curve D can be computed by integration.

$$\frac{dr_e}{d\theta} = \frac{\dot{r}_e}{\dot{\theta}}. \tag{15.17}$$

At the BUP points $\theta_{bup} = \sin^{-1}\left(\frac{1}{w}\right)$ we have that $\dot{r_e} = 0$ and also $\dot{r} = 0$ so the slope of curve D at its initial point at the BUP should be

$$\frac{dr_e}{d\theta}(\theta_{bup}) = 0.$$

Let us proceed now to the construction of an *envelope barrier* emanating from the curve D. To see if this is feasible we have to check if the following two problems have unique solution.

1.
$$\begin{array}{l}\min\left[\bar{n}_\theta \dot{\theta}(\varphi, \tilde{\psi}(\varphi)) + \bar{n}_{r_e} \dot{r}_e(\varphi, \tilde{\psi}(\varphi))\right],\\ \varphi, \text{ where } \ w\cos\tilde{\psi}(\varphi) = \cos\varphi,\end{array} \tag{15.18}$$

2.
$$\max_{\psi}\left[n_r \dot{r}(\overset{*}{\varphi}, \psi) + n_\theta \dot{\theta}(\overset{*}{\varphi}, \psi) + n_{r_e} \dot{r}_e(\overset{*}{\varphi}, \psi)\right). \tag{15.19}$$

It is easy to see that the first problem has a unique solution for φ (corresponding to trajectories that do not arrive at $r = \rho$ from the region $r < \rho$). So

$$\overset{*}{\varphi} = \tilde{\varphi}. \tag{15.20}$$

The second problem has also a unique solution so

$$\overset{*}{\psi} = \tilde{\psi}(\overset{*}{\varphi}) = \tilde{\psi}(\tilde{\varphi}). \tag{15.21}$$

Relation (15.20) and relation (15.21) mean that at each point of the curve D the 2-dimensional *barrier controls* $(\tilde{\varphi}, \tilde{\psi}(\tilde{\varphi}))$ that generate the curve D as a semipermeable surface with respect to a motion constrained to $r = \rho$ and the 3-dimensional *barrier controls* $(\overset{*}{\varphi}, \overset{*}{\psi})$ that generate the surface B as a manifold of *barrier paths* in the game-set have identical components. This knowledge enables us to construct this manifold and obtain an *envelope barrier* emanting from the BUP.
In our problem this is best done in the *real frame* where the *regular barrier paths* correspond to motions of both players along straight lines.

•

Bibliography

[ABB65] Y.C. Ho A. Bryson and S. Barron. Differential games and optimal pursuit evasion strategies. *IEEE Trans. on Automatic Control*, ac-10:385–389, 1965.

[ABL69] A. Gerard A. Blaquiere and G. Leitman. *Qualitative and Quantitative Games*. Academic Press, New York, 1969.

[Bas90] T. Basar. Game Theory and $\mathcal{H}^\infty$-Optimal Control. In *Differential Games Developments in modelling and computations*, Springer Verlag, Berlin, 1990.

[BB91] T. Basar and P. Bernhard. *$\mathcal{H}^\infty$-Optimal Control and Related Minimax Design Problems*. Birkhauser, Boston, 1991.

[Ber64] L.D. Berkovitz. A variational approach to differential games. *Ann. Math. Study*, 52:127–174, 1964.

[Ber75] P. Bernhard. New results about corners in differential games, including state constrasints. In *Proceedings 6th IFAC World Congress*, Cambridge, Mass., 1975.

[Ber77] P. Bernhard. Singular surfaces in differential games. In H. W. Knobloch P. Hagedorn and G. J. Olsder, editors, *Differential Games and Applications*, Springer-Verlag, 1977.

[Ber79] P. Bernhard. Linear quadratic, two person zero sum differential games; necessary and sufficient conditions. *Journal of Optimization Theory and Applications*, 27:51–69, 1979.

[Ber87] P. Bernhard. Differential games: isaacs' equation. In Madan Singh, editor, *Encyclopedia of Systems and Control*, Pergamon Press, 1987.

[Bre] J.V. Breakwell. Sufficient conditions in zero sum differential games. preprint.

[Bre75] J.V. Breakwell. Pursuit of a faster evader. In J. D. Grote, editor, *The Theory and Application of Differential Games*, D. Reidel Publishing Company, Dordecht, Holland, 1975.

[Bre89] J.V. Breakwell. Time optimal pursuit inside a circle. In T. S. Basar and P. Bernhard, editors, *Differential Games and Applications*, Springer-Verlag, Berlin, Germany, 1989.

[BS89] M. Bardi and P. Soravia. A pde framework for games of pursuit evasion type. In T.S. Basar and P. Bernhard, editors, *Differential Games and Applications*, Springer-Verlag, Berlin, 1989.

[Chi76] S.A. Chigir. The game problem on the dolichobrachistochrone. *PMM*, 30:1003–1013, 1976.

[Cil70] M.D. Ciletti. On the contradiction of bang-bang-bang surface in differential games. *Journal of Optimization Theory and Applications*, 5, 1970.

[DS85] A. Davidovitz and J. Shinar. Eccentric two target model for qualitative air combat analysis. *AIAA Journal of Guidance, Control, and Dynamics*, 1985.

[Fly73] J. Flynn. Lion and man: the the boundary constraint. *SIAM Journal on Control*, 11:397–411, 1973.

[Fly74] J. Flynn. Lion and man: the general case. *SIAM Journal on Control*, 12:581–597, 1974.

[Fri72] A. Friedman. Comparison theorems for differential games. *Journal of Differential Equations*, 12:162–172, 1972.

[GL75] S. Gutman and G. Lietman. On a class of linear differential games. *Journal of Optimization Theory and Applications*, 17:511–522, 1975.

[GL76] S. Gutman and G. Lietman. Optimal strategies in the neighbourhood of a collision course. *AIAA Journal*, 14:1210–1212, 1976.

[Ho65] L. Ho. Review of the book differential games, by r. isaacs. *IEEE Trans. on Automatic Control*, 1965.

[Isa69] R. Isaacs. Differential games: their scope, nature and future. *Journal of Optimization Theory and Applications*, 3:283–295, 1969.

[Isa75] R. Isaacs. *Differential Games.* Robert E. Krieger, New York, 1975.

[Isb67] J.R. Isbell. Pursuit around a hole. *Naval Research Quarterly*, 14, 1967.

[JDF89] P.P. Khargonekar J.C. Doyle, K. Glover and B.A. Francis. StateState-Space Solutions to Standard $\mathcal{H}_2$ and $\mathcal{H}_\infty$ Control Problems. *IEEE Transactions on Automatic Control*, AC-34:831–847, 1989.

[LB75] J. Lewin and J.V. Breakwell. The surveillance-evasion game of degree. *Journal of Optimization Theory and Applications*, 16:339–353, 1975.

[Les85] D. Leshem. *Composite Barriers and Corner Conditions in Differential Games.* PhD thesis, Stanford University, 1985.

[Lew76] J. Lewin. The bang-bang-bang problem revisited. *Journal of Optimization Theory and Applications*, 18:429–432, 1976.

[Lew86] J. Lewin. The lion and man problem revisited. *Journal of Optimization Theory and Applications*, 49:411–430, 1986.

[LO79] J. Lewin and G.J. Olsder. Conic surveillance evasion. *Journal of Optimization Theory and Applications*, 27:107–125, 1979.

[Ls85] P.L. Lions and P.E. Souganidis. Differential games, optimal control and directional derivatives of viscosity solutions of bellman's and isaacs' equations. *SIAM Journal on Control*, 23:566–583, 1985.

[Mag76] E.F. Mageirou. Values and strategies for infinite time linear quadratic games. *IEEE Transactions on Automatic Control*, ac-21:547–550, 1976.

[Mer71] A.W. Merz. *Homicidal Chauffeur, A Differential Game.* Sudaar, Guidance and Control Laboratory Report 418, Stanford University, 1971.

[MH77] E.F. Mageirou and Y.C. Ho. Decentralized stabilization via game theoretic methods. *Automatica*, 13:393–399, 1977.

[SR91] J.L. Speyer and I. Rhee. A game theoretic approach to a finite-time disturbance attenuation problem. *IEEE Transactions on Automatic Control*, AC-36:1021–1032, 1991.

[Tay70] J.G. Taylor. *Application of Differential Game Theory to Problems of Naval Warfare: Surveillance Evasion.* AD report 717587, Naval Academy, 1970.

[UF90] K. Uchida and M. Fujita. Controllers attenuating disturbances and initial uncertainties for time varying systems. In *Differential Games Developments in modelling and computations*, Springer Verlag, Berlin, 1990.

[Wil72] D.J. Wilson. On a nonsingular surface of a differential game. *Journal of Optimization Theory and Applications*, 9, 1972.

Index